Executive's Cybersecurity Program Handbook

A comprehensive guide to building and operationalizing a complete cybersecurity program

Jason Brown

BIRMINGHAM—MUMBAI

Executive's Cybersecurity Program Handbook

Group Product Manager: Mohd Riyan Khan

Publishing Product Manager: Prachi Sawant

Senior Editor: Tanya D'cruz

Technical Editor: Shruthi Shetty

Copy Editor: Safis Editing

Project Coordinator: Deeksha Thakkar

Proofreader: Safis Editing

Indexer: Pratik Shirodkar

Production Designer: Alishon Mendonca

Marketing Coordinator: Marylou De Mello

First published: February 2023

Production reference: 1270123

Published by Packt Publishing Ltd.

Livery Place

35 Livery Street

Birmingham

B3 2PB, UK.

ISBN 978-1-80461-923-0

www.packtpub.com

To my wife and my daughter. Words cannot express the love I have for both of you. I am so thankful to experience the love and joy you bring me every day. I love you both.

– Jason Brown

Contributors

About the author

Jason Brown's passion lies in data privacy and cybersecurity. He has spent his career working with businesses, from small to large international companies, developing robust data privacy and cybersecurity programs. Jason has held titles such as chief information security officer, virtual chief information security officer, and data privacy officer. He has obtained many industry-leading certifications, including ISC2's CISSP, ISACA's CDPSE and COBIT, and ITIL, and holds a Bachelor of Science degree from Central Michigan University and a Master of Science degree from Ferris State University.

About the reviewers

Sai Praveen Kumar Jalasutram is an experienced cybersecurity leader on a mission to defend organizations from advanced cyber threats. With a strong track record of leading teams in conducting investigations, gathering intelligence, implementing security controls, and improving the overall security posture, Sai has a deep understanding of the constantly evolving threat landscape and is skilled in identifying threat patterns and developing strategies to mitigate potential cyber threats. Throughout his career, Sai has worked with a variety of global organizations across multiple sectors, including government, technology, finance, and education, consistently delivering valuable insights and solutions for improving the cyber resilience of these companies. Sai played a key role in the development of the **Certified Ethical Hacker (CEHv10)** and **Certified Threat Intelligence Analyst (CTIA)** certifications offered by EC-Council and has published articles in various cybersecurity magazines.

Kenneth Underhill is a co-founder of Cyber Life, which provides cybersecurity certification education to organizations and cybersecurity professionals, and one of the authors of the international bestselling book *Hack the Cybersecurity Interview*. He has won multiple industry awards for his work to improve diversity in the industry and is an advocate for women's rights. Ken educates millions of people each year through his online cybersecurity courses and sits on the advisory board of **Breaking Barriers Women in CyberSecurity (BBWIC)**. He also holds a graduate degree in cybersecurity and information assurance and several industry certificates (CCSK, CEH, and CHFI).

Table of Contents

3

Cybersecurity Strategic Planning through the Assessment Process 25

Part 2 – Administrative Cybersecurity Controls

4

Establishing Governance through Policy 39

5

The Security Team 53

6

Risk Management   67

7

Incident Response   87

Part 3 – Technical Controls

9

10

11

Securing Software Development through DevSecOps   163

12

Testing Your Security and Building Metrics   183

Preface

The *Fog of More* is a term created for the onslaught of vendors and cybersecurity professionals who want to throw money at a problem. The problem turns into appliances, with blinky lights, that never get fully implemented or go unused. Hundreds of thousands, if not millions, of dollars are spent every year on buying the latest and greatest technology only to miss the fundamentals of implementing basic security controls. We, as professionals, need to take a step back and look at the overall current state of the cybersecurity program and decide where we want to go. This can get muddled with wanting to spend money on technologies, but is that truly the answer?

Many cybersecurity professionals who are expected to develop a cybersecurity program begin with little to no budget, with the expectation of protecting the organization from unforeseen events. While some technologies cost money, cybersecurity fundamentals still remain. These fundamentals are oftentimes overlooked, spending more money to compensate for controls that may already be built into existing resources. There has to be a better way to tackle this problem.

This book will guide you through low-cost/no-cost solutions that you can start off with when creating a cybersecurity program from the ground up. There are plenty of ways that you can develop a program when you do not have the initial funding to get the program going. Utilizing cybersecurity standards is just one way of getting there, guiding you through the process of implementing best practices. These all have to be planned out by setting goals to achieve the objectives in your strategic plan.

Who this book is for

You have climbed the ranks and now it is your turn to lead a team of cybersecurity rockstars. As the new head of security, it is time to leave most of your technical abilities behind and focus on the business. This book will help you change your mindset by taking business objectives and applying them to the cybersecurity outcomes the organization needs. This book is meant for the first-time manager, director, or chief information security officer – those who have climbed the ranks and are now in charge.

What this book covers

Chapter 1, The First 90 Days

Congratulations! You are now the head of security for your organization – what a tremendous achievement. If you have worked your way up through the ranks of IT and cybersecurity, you have had to develop the technological side of your career. It is not a time to abandon this, however; you will need to understand how the business runs and what is important to the organization. In addition to understanding your business goals, in this chapter, we also review and develop strategies for determining which IT and security aspects are important to you and your team.

Chapter 2, Choosing the Right Cybersecurity Framework

Cybersecurity is more than just "checking a box," it is a mindset. It is a continuously evolving field with new threats occurring every day. How do you stay ahead of the adversary? In order to stay a step ahead, we will review the various types of cybersecurity frameworks available and select one to implement. Cybersecurity frameworks promote best practices, which in turn will make it more difficult for an adversary to take control of an IT resource.

Chapter 3, Cybersecurity Strategic Planning Through the Assessment Process

Once we have picked our framework of choice, we must put it to work. Where do you start? The first step is to conduct a risk assessment and perform a gap analysis. This analysis will not only show you your current state of security but will also show you where your deficiencies are. From there, we will create our strategic plan to put safeguards in place to enhance the security of the environment.

Chapter 4, Establishing Governance Through Policy

A large part of a maturity-based cybersecurity program is the establishment of governance. Governance is an important step in maturing your cybersecurity, or overall IT, department. Part of your governance program is developing policies, standards, and procedures for your teams to align with. There also must be a framework for establishing policy documents for your organization. Creating them is the first step; however, you must develop a document life cycle for review and approval. This chapter will review the steps necessary to create a governance program.

Chapter 5, The Security Team

As the head of security, you may have a large team, have a handful of employees, or you might be the lone person working on security. Eventually, you will want to continue to grow your team, but what should you look for? Cybersecurity analysts, engineers, and architects all have their own sets of unique qualities, so how do you choose which is right to meet the business' demanding requirements? This chapter will provide an overview of these positions, and ensure that you have the backing of the business to continue to mature your department.

Chapter 6, Risk Management

There is nothing that comes without some type of risk and these risks are all driven by decision-making. This is no different when evaluating risks within an IT environment. In this chapter, we review how to use security categorizations to evaluate the risk for an IT resource. Having evaluated it, we continue to look at how we want to maneuver that risk – whether we accept, transfer, mitigate, or avoid it. Once we have determined how we want to maneuver the risk, we will build out documentation through the use of a systems security plan and risk register to record the risk posture of the environment.

Chapter 7, Incident Response

Some aspect of your IT and security program will fail. Someone will penetrate your network, or a catastrophic event will occur. You must train as you fight, and doing so will better prepare you for when the inevitable happens. This chapter reviews how to prepare for an incident, what to have in place for when one occurs, and how to better prepare yourself for when it happens again.

Chapter 8, Security Awareness and Training

This chapter reviews the differences between security awareness, training, and education. It will walk you through how to gain buy-in from executives to train your security staff. As cybersecurity continuously evolves, so too should the knowledge of your employees.

Chapter 9, Network Security

There have been plenty of digital disruptions throughout history, though by far the largest disrupter has been the creation of the internet. There are approximately 30 billion devices all connected to the same medium, which can be challenging to secure, to say the least. In this chapter, we will review the history of the internet, predict what the next biggest thing will be in internet and networking technologies, and discuss how to secure the next invention.

Chapter 10, Computer and Server Security

Ever wondered how computers got their start? This chapter reviews the history of computers, server architecture, and the software that runs on them. We will look at how to secure them and mitigate any threats through best practices.

Chapter 11, Securing Software Development Through DevSecOps

Software is what drives the technologies we use every day. Through software, we can write applications, create operating systems, and direct traffic across a network. It is no wonder software development is key to any business. However, it can also be our downfall. This chapter reviews how to implement security early in a project, the DevSecOps steps, the importance of code reviews, and how to test software to ensure that it has mitigated any risks to a tolerable level.

Chapter 12, Testing Your Security and Building Metrics

You have put in the work, and now it is time to show how your security program has progressed. As the head of security, you will need to highlight your achievements and show the executive team and the board what work has been done and what is yet to be done.

Download the color images

We also provide a PDF file that has color images of the screenshots and diagrams used in this book. You can download it here: `https://packt.link/QTWim`.

Get in touch

Feedback from our readers is always welcome.

General feedback: If you have questions about any aspect of this book, email us at `customercare@packtpub.com` and mention the book title in the subject of your message.

Errata: Although we have taken every care to ensure the accuracy of our content, mistakes do happen. If you have found a mistake in this book, we would be grateful if you would report this to us. Please visit `www.packtpub.com/support/errata` and fill in the form.

Piracy: If you come across any illegal copies of our works in any form on the internet, we would be grateful if you would provide us with the location address or website name. Please contact us at `copyright@packt.com` with a link to the material.

If you are interested in becoming an author: If there is a topic that you have expertise in and you are interested in either writing or contributing to a book, please visit `authors.packtpub.com`.

Share Your Thoughts

Once you've read *Executive's Cybersecurity Program Handbook*, we'd love to hear your thoughts! Scan the QR code below to go straight to the Amazon review page for this book and share your feedback.

`https://packt.link/r/180461923X`

Your review is important to us and the tech community and will help us make sure we're delivering excellent quality content.

Download a free PDF copy of this book

Thanks for purchasing this book!

Do you like to read on the go but are unable to carry your print books everywhere?

Is your eBook purchase not compatible with the device of your choice?

Don't worry, now with every Packt book you get a DRM-free PDF version of that book at no cost.

Read anywhere, any place, on any device. Search, copy, and paste code from your favorite technical books directly into your application.

The perks don't stop there, you can get exclusive access to discounts, newsletters, and great free content in your inbox daily

Follow these simple steps to get the benefits:

1. Scan the QR code or visit the link below

https://packt.link/free-ebook/9781804619230

2. Submit your proof of purchase

3. That's it! We'll send your free PDF and other benefits to your email directly

Part 1 – Getting Your Program Off the Ground

All too often I see organizations throwing hundreds, if not thousands of dollars, at cybersecurity problems. But is this really the right course of action, especially when you are only just starting to develop your program?

To start your endeavor in developing a cybersecurity program, you must understand where you are starting from. Often, this will require an assessment to be performed, one that helps you understand the current security posture. To perform an assessment, you must first choose the types of controls and cybersecurity standard you want to assess the organization against.

There are plenty of cybersecurity standards to choose from, but how do you choose? First, it requires you to understand the organization, its requirements, and where it conducts business. If it conducts business in Europe or other regions outside the U.S., it may be best to use an international standard such as ISO 27001/27002. In the U.S., many higher education and governmental organizations typically use standards developed by the **National Institute of Standards and Technology (NIST)**.

This first part of the book is centered around cybersecurity program development, and how you and your organization can leverage the NIST standards. The first few chapters will guide you through what you need to do in the first 90 days on the job, which standard you should choose, and how to align that standard to business objectives. By leveraging the NIST Cybersecurity Framework, you can get a head-start on the implementation of administrative controls and establish governance. In addition to administrative controls, we will also look at the **Center for Internet Security**, a U.S. non-profit organization that assists with the implementation of technical security controls. Combining these two frameworks together will assist in the creation of your cybersecurity program.

1
The First 90 Days

Congratulations and welcome to the cybersecurity club! Whether you are just starting off on your cybersecurity career or are a seasoned professional, your first 90 days as the head of security can be tough – yet rewarding. During your first 90 days as the head of security, you will be challenged to learn the business and its processes, build new relationships, and gain an understanding of what is important to the company. You will need to hold meetings with your peers to better understand the technology or security stacks being used at the company.

The ability to develop strong relationships early on in your tenure will pay off in the long run. Your co-workers will get to not only know you by name but also get to know you personally and professionally. This is also your opportunity to do the same. The key is to build good, strong relationships early so you are not only approachable, but people feel comfortable talking to you.

Many in the business world see information security as the department of, *"No!"* where security trumps everything, from stopping projects to blocking new business processes. Information security departments must take a different stance when it comes to dealing with kids running with scissors. We must engrain ourselves into the business to see how it runs, and that will come through relationship building. In this first chapter, we will begin to look at what steps you should take in the first 90 days as the head of security at your organization.

In this chapter, we'll be covering the following topics:

- Getting executive buy-in
- Budget or no budget?
- Vision statements
- Mission statements
- Program charters
- The pillars of your cybersecurity program

Getting executive buy-in

Building relationships with your peers is a must, and this includes those on the executive team. You want to build the same rapport with your executive team as with your peers, so they too feel comfortable speaking with you. This is also the time to begin discussing their thoughts about what the security program was meant to do and what the original direction was. These discussions do not stop with the executive team; if you are able, have the same discussion with key stakeholders and the board of directors.

Without getting executive buy-in, your program may stall or go nowhere. There are many reasons for this; however, the first step is to determine what the business needs are and how executives see them being achieved. Many professionals want to come in and inflict change in the cybersecurity program right away – I would advocate against this, at least for a little while. The reason is that you must understand what is important to the business. Remember, you have to crawl before you can walk. Information technology and cybersecurity are no different in their approaches when determining the vision you and other executives have of the cybersecurity program.

Cost and budget are also key components of the program, but are not as important as getting executive buy-in. A new head of security or chief information security officer could come in, build out the plan, forecast the budget, utilize several free open source cybersecurity tools, and it could still go nowhere as the executive team has decided to not move forward. It did not stop because of the budget – that was just a component of it. The real issue is that your executive team does not accept the decisions or technology planned for implementation. This is why getting executive buy-in and having conversations regarding the cybersecurity program are so important.

Next, we can begin to build out a budget that makes sense to everyone. Whether it is five dollars or a million dollars, there are plenty of ways to secure the business; however, the direction you want to go in with the program is up to you.

Budget or no budget?

A budget will make or break a department. Change my mind! That is a very true statement. Or is it? It is what you do with the budget you are given that will make or break a department. Organizations are hard-pressed to spend money on cybersecurity. Why? Because people think "It has never happened to us yet, so why should we bother?" or "We are too small to be a target." However, mindsets are beginning to change. Many on the business side see cybersecurity or information technology as a sinkhole – one the business pours money into but never sees anything come of it. It is up to you to sell your ideas, get funding, and spend it accordingly.

Cybersecurity, however, is not something to take lightly or brush off as a second thought. States and the federal government are enacting breach notification laws for public and private sector organizations. Cyber insurance companies are serious, too, as they want to ensure their clients are performing their due diligence in reducing cyber risk across the organization. While not everything requires spending the business's hard-earned money, certain aspects of cybersecurity do require funding.

Cybersecurity spending is also reaching all-time highs. Between 2021 and 2025, cybersecurity spending is expected to reach $1.75 trillion. According to Steve Morgan, founder of Cybersecurity Ventures, the market was only worth $3.5 billion in 2004 (Braue)[1]. That is a 500% increase (I am terrible at math). *The Fog of More* is a phrase characterized in cybersecurity as vendors trying to sell you the newest, shiniest blinky green and amber lights. How many of us have purchased new firewalls, only for them to collect dust? Better yet, how many of us have implemented firewalls with *allow any/any/any rules*? Don't believe me? Performing a quick search on Shodan provides some disturbing statistics. For instance, there were over 286,000 results for open Telnet ports; VNC, almost 540,000 results; RDP, 582,000. These protocols, when exposed to the internet, increase the risk of your organization being attacked.

Not all vendors are bad; quite a few legitimately want to help. In the end, however, money talks. This is why it makes sense to rephrase "A budget will make or break your department " to " it is what you do with the budget that will define the department." Do not let others fool you. You can churn out a robust security program without major funding.

Organizational technology stacks come in two different flavors, build and buy. While the *build* camp prefers to utilize as much open source technology and free utilities as possible, the *buy* camp wants to ensure that not only do they have support from a company, but they also have a single entity to blame if something goes wrong. While some concepts in this book will require a company to purchase some type of IT resource, this book is not centered on CapEx purchases. There are plenty of free, low-cost, or no-cost solutions out there if your team is willing to allocate the resources and time to learn about concepts and learn the skills of the trade.

For instance, begin developing your company's vision, mission, and program charters. This will set the foundation for your program.

Vision statements

The vision of the office of information security is to secure the organization while making security a second thought.

Many organizations tend to throw technology at a problem, but is that the right solution? What is the goal of the company's information security program? What will make you and your team stand out as a force for delivering top-notch security services? As a security leader, you must first understand where you are and where you want to go. If you do not have an end goal, how will you know how to get there?

A vision statement is a high-level description of how the program strives to achieve success. For instance, the preceding quote is a vision statement that could be used for a security department. It is intended to not only state the purpose of the department but the overall goal. A phrase I like to use for our security program is "Employees already think of cybersecurity as a second thought – I intend to keep it that way."

Why is that statement important to me and our program? We want our security program to be as robust as possible and protect our systems and data while keeping our users safe. Information security should enable while making it easy for those who are not technically savvy. It should be as transparent as possible without always being in your face. Users should not have to read an entire manual to learn how to do their jobs, which are already tough without adding more layers on top.

The vision statement should depict what is most important to the department or organization. It should not be lengthy—only three sentences or fewer, but make it meaningful. It can be internal or external customer-facing, but make it a way of marketing yourself to others. As the security field is dynamic, a vision statement does not have to remain static and can evolve over time. One could write a vision statement and a few years down the line, decide to change it.

Here's an example of a vision statement:

> *The Institute for Information Security & Privacy (IISP) at Georgia Tech is as an international leader in researching, developing, and disseminating technical solutions and policy about cybersecurity and privacy. We assemble strong, innovative, multi-disciplinary teams to address contemporary and future cybersecurity or privacy challenges faced by government, industry and individuals. Our graduates become leaders in government, scientific, industry and entrepreneurial communities.*
>
> —*Georgia Tech University* (`https://www.scs.gatech.edu/ research/institutes-centers`)

There is no right way or wrong way to create a vision statement for your department. With one in place, however, it provides context for the goals and objectives that the department strives to achieve. It also shows that the department takes cybersecurity seriously in the types of services it will provide to its customers.

While vision statements are important for providing context for what the department strives to achieve, mission statements are equally as important. Mission statements depict why the department exists.

Mission statements

One may think, "Doesn't a mission statement belong to the business?" While businesses have a mission statement, departments should have one too. Business-style mission statements articulate the purpose of the business/department or establishes the reason for their existence. Some famous mission statements include the following:

> *To bring the best user experience to its customers through its innovative hardware, software, and services.*
>
> —*Apple* (`https://mission-statement.com/apple/`)

To empower every person and every organization on the planet to achieve more.

—*Microsoft (*`https://www.comparably.com/companies/`
`microsoft/mission`*)*

Accelerating the world's transition to sustainable energy.

—*Tesla (*`https://www.tesla.com/about`*)*

What is your security department's reason for existing? Is it to protect your organization's sensitive data? Is it to thwart those who would do harm to your organization? How about securing and protecting the free flow of information across the world? I am sure we all have a back story as to why the newly minted cybersecurity manager or chief information security officer position opened at the company. What were the circumstances around your position being created? These questions do not necessarily have or need to have answers. They are there to help you decide how to construct a mission statement.

Before writing a mission statement, understand the culture and, again, what is important to the business. Write one, or a few, select the ones you like best, and then solicit feedback. If you are lucky and have a few security employees that work with you, get their feedback too or ask them to join in.

Mission statements, much like vision statements, are not long: maybe a sentence or two. However, it must have meaning and be celebrated as the crux of how the department will operate. It should motivate employees to do better and be better. The statement should also cause your customer base (others inside and outside the organization) to want to contact you when something is wrong—or right! On day one, a new employee should know and understand the department's mission statement.

Mission and vision statements are great in that they highlight the department's importance and reasons for being a crucial part of the organization. Program charters help bring everything together as they show what a department is responsible for.

Program charters

What is your department responsible for? How will you go about setting policies for information technology and the rest of the organization? Does the department have oversight of how things are implemented, configured, and monitored? When establishing governance, the first thing people think of is building roles and responsibility matrices – **responsible, accountable, consulted, and informed (RACI)** charts, and the like.

While RACI charts, roles, and responsibility matrices tend to provide the *who* and the *what*, they do not provide much detail. Program charters are intended to help fill in those gaps. They can be written to provide as little or as much detail as possible to help define what those responsibilities are and their intended purpose. For instance, most information security departments act as advisors for the rest of the information technology department. In this scenario, security has oversight of all aspects of information technology, but security does not implement or configure the IT resource (because of the separation of duties).

Much like policies, standards, and procedures (which we will cover in *Chapter 4*), a program charter document should have the following sections:

Purpose

What is the purpose of the charter? What is it trying to convey to the reader? Are there specific questions that the charter is trying to answer?

Scope

Charters impact an organization in many ways. They can impact internal and external employees, third-party vendors, contractors, whole departments, or the organization. Who will be impacted when this charter is put into place?

Responsibilities

Much like a RACI chart, what is the security department instructed to do? What is it responsible for? This is where you set the stage for how the department will function. Will it have oversight of many different aspects of information technology and the rest of the organization?

Those responsible for the charter

The charter must have a stakeholder and an executive sponsor to sign off on it. The stakeholder should be the head of the department, whether that is the **Chief Security Officer (CSO)**, **Chief Information Security Officer (CISO)**, director of information security, or manager of information security. These are the individuals who will be making decisions about how they see their cybersecurity department operating. The executive sponsor, whether that is a **Chief Information Officer (CIO)** or **Chief Executive Officer (CEO)**, must have the authority to sign off on the charter. Once the charter is officially signed off on, it will have the teeth necessary to carry out the charter and any other supporting documentation.

In the previous sections, we have discussed how the security department will achieve success, its importance to the organization, and what it will be responsible for. To build on those concepts, we will take it a step further to discuss initiatives, strategy, and what is important to you as a leader.

The pillars of your cybersecurity program

What are the key initiatives for your security program? What is the strategy you will set that will direct the security program over the next 3-5 years? While we will talk more about developing a strategy in the next two chapters, this is where we will set the stage for that vision. First, start off with two to five high-level categories that are important to you and then begin to drill down from there. Each subcategory gets more defined as we drill down. An example of this is found in the following graphic:

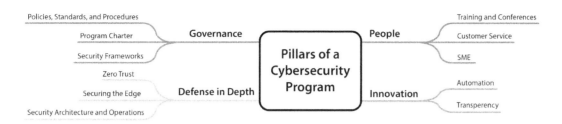

Figure 1.1 – Cybersecurity pillars

In *Figure 1.1*, the diagram depicts what could be important for your security program. Major high-level ideas spur off from the main topic: **Governance**, **Defense In Depth**, **People**, and **Innovation**. From the high-level categories, it begins to narrow down from strategic to tactical. For instance, **Innovation** drills down to **Automation** or **Transparency** and **Defense in Depth** drills down to **Zero Trust**, **Securing the Edge**, and **Security Architecture and Operations**.

The subcategories begin to drill down into specific concepts or technologies, but do not state which technologies are used. You should not have more than four subcategories from the main topic, but you can have as many or as few as needed to tell the story. This begins the development of the strategic vision you have for your department.

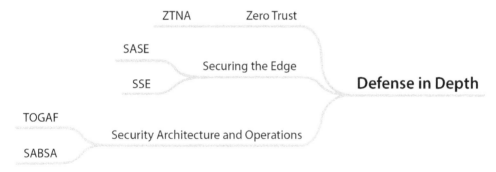

Figure 1.2 – Cybersecurity Defense in Depth pillar

The preceding figure depicts an expanded view of what defense in depth could look like. Starting with the main topic, **Defense in Depth**, we move toward the categories **Zero Trust**, **Securing the Edge**, and **Security Architecture and Operations**, followed by their respective subcategories.

Summary

Your first 90 days as a new manager, director, or CSO can be an exciting yet intimidating time. How will you get a budget for your program? How does executive management see information security? How will you develop a cybersecurity strategy? There are so many questions, yet few initial answers. Many come into the position and begin throwing technology at the problem, drowning in the *Fog of More*.

Remember, a security program is more than just technology; it also consists of people and processes. The key to getting started during your first 90 days is to understand the business, its processes, and how key stakeholders see the alignment of information technology and security to the business. Begin developing the department's mission and vision statements and evangelize them throughout. Get others involved when creating these documents to gather their input and see what is important to them too.

As you have learned in this chapter, your first 90 days is also a time for creating new relationships with your coworkers. Relationships matter when it comes to information technology and security – make sure they are a priority. Eventually, you will have to work with colleagues from all different aspects of the business. Coworkers from finance, human resources, manufacturing, and other departments will have to be incorporated into your processes. Business continuity, incident response, and risk management are not information technology problems; they are business problems. As such, personnel from these departments need to be involved too.

We have now set the initial structure for the cybersecurity department, what is important, and why. In the next chapter, we will discuss the importance of a cybersecurity framework and its overall impact on the department and the organization. These two chapters help set the foundation you can begin to build a strategy on moving forward.

References

1. Braue, David. (2021, September 10). *Global Cybersecurity Spending to Exceed $1.75 Trillion From 2021-2025*. Cyber Crime Magazine. `https://cybersecurityventures.com/cybersecurity-spending-2021-2025/`

2
Choosing the Right Cybersecurity Framework

The evolving technological landscape has significantly impacted business operations, public institutions, critical infrastructure, and how individuals interact with each other. The accessibility of information and exponential growth in the number of internet-connected devices contribute to a worldwide proliferation of cybersecurity threats. These threats have impacted hospitals, rendering them incapable of accepting patients due to infections of ransomware, and caused a temporary shutdown of municipal transportation, system-wide failures in the stock market, and widespread internet outages.

You will encounter many things throughout your journey as a cybersecurity professional. From the start, building a new cybersecurity program can be difficult, or at least no one will say it is easy. There are tools widely available to help you begin your journey. Cybersecurity frameworks help answer the questions you may have such as where to start or how to implement technology. Most cybersecurity frameworks you will encounter are available for free, and some for a nominal cost. Whatever framework you choose, the chapter will help direct you in the right direction in deciding on a cybersecurity framework.

In a nutshell, we'll be covering the following topics in this chapter:

- What is a cybersecurity framework?
- Security as a checkbox
- Continual improvement
- Selecting the right framework
- The framework used in this book

What is a cybersecurity framework?

A framework provides the who, what, where, why, and how of implementing security controls and best practices. Some might decide to implement the entire framework, while others may take bits and pieces from certain sections. In other circumstances, maybe you want to expand on implementing controls to apply additional security functionality. Think of a home blueprint. A new home buyer may want to use just the default home layout like everyone in a subdivision. Others may want the default layout but with some additions to it, such as expanding the bathroom for a hot tub or a double vanity. Maybe a new home buyer wants a standard home, with a white picket fence, but without landscaping. These are all examples, but I think you get it.

As you will learn later in this chapter, in the version 8 release of the **Center for Internet Security Critical Security Controls** (**CIS CSC**), there are categories of controls or **implementation groups** (**IGs**). These controls are ranked based on the size of the IT and security departments. For instance, the CIS controls state that for small to medium-sized organizations, tracking authorized and unauthorized devices on your network is a necessity, whereas **dynamic host configuration protocol** (**DHCP**) logging to automatically update the database is meant for larger, more mature organizations.

There are hundreds of different cybersecurity frameworks out there; how do you choose? First, speak with organizational stakeholders to find out what regulatory requirements the business must meet. If you store or transmit credit card data, then you are subjected to the **Payment Card Industry Data Security Standards** (**PCI-DSS**). If you hold on to certain types of U.S. government data, then you may be subjected to the **Federal Information Security Management Act** (**FISMA**) or the **Defense Federal Acquisition Regulation Supplement/Cybersecurity Maturity Model Certification** (**DFARS/CMMC**). The business may also be subjected to privacy laws and regulations such as the European Union's **General Data Protection Regulation** (**GDPR**), which regulates data transfers across international borders. As a new cybersecurity professional, you must know whether your business is subjected to these and others.

Types of cybersecurity frameworks

There are many types of cybersecurity frameworks to choose from, but how do you know whether you are picking the right one? Before selecting or building your own cybersecurity framework, you must first understand the business goals and objectives. You must know whether the organization conducts business locally or internationally. The following is an overview of the various standards that can be used. A seasoned professional can also create their own – in fact, many do. By taking what you deem necessary for your organization, you can take parts of different cybersecurity standards and mesh them together. The following will provide an overview of various frameworks that you can use to help bolster your program.

NIST

Nestled inside the Department of Commerce, the **National Institute for Standards and Technology** (**NIST**) has been directed to develop industry standards, including information technology and cybersecurity. Founded in 1901, it is one of the oldest departments within the United States federal government. NIST has developed four different types of technical publications, including the following:

- **Federal Information Processing Standards (FIPS)**
- **Special Publications (SP)**
- **NIST Internal or Interagency Reports (NISTIR)**
- **NIST Information Technology Laboratory Bulletins (ITL Bulletins)**

In the next few subsections, we will look at some common NIST standards you may encounter on the job.

SP 800-53 Revision 5

SP 800-53 is probably the best-known security framework in the United States. SP 800-53 consists of 20 privacy and security control families with an estimated 1,000 individual controls. Information systems used by U.S. federal government agencies are required to follow this security standard in order to meet FISMA requirements. The control families consist of the following:

ID	Family	ID	Family
AC	Access Control	PE	Physical and Environmental Protection
AT	Awareness and Training	PL	Planning
AU	Audit and Accountability	PM	Program Management
CA	Assessment, Authorization, and Monitoring	PS	Personnel Security
CM	Configuration Management	PT	Personal Identifiable Information (**PII**) Processing and Transparency
CP	Contingency Planning	RA	Risk Assessment
IA	Identification and Authentication	SA	System and Services Acquisition
IR	Incident Response	SC	System and Communications Protection
MA	Maintenance	SI	System and Information Integrity
MP	Media Protection	SR	Supply Chain Risk Management

Table 2.1 – NIST 800-53 control families

Before you can begin with SP 800-53 , you must know where to start by understanding the risk of the asset and the information that resides on it. To calculate the overall risk of the asset, NIST developed FIPS 199. The document assists agencies and businesses alike to better understand and quantify risk. With risk ratings of low, moderate, and high, you take the highest watermark and apply the controls from SP 800-53. The risk ratings are as follows:

- **Low**: The risk to an information system is low if the loss of confidentiality, integrity, or availability has limited or no adverse effects on the organization.

- **Moderate**: The risk to an information system is moderate if the loss of confidentiality, integrity, or availability has serious adverse effects on the organization.

- **High**: The risk to an information system is high if the loss of confidentiality, integrity, or availability has severe or catastrophic adverse effects on the organization.

To calculate the impact on an information system according to FIPS 199, use the following equation:

$$SC = \{(confidentiality, impact), (integrity, impact), (availability, impact)\}$$

In this equation, *SC* stands for security category, and *impact* represents the potential impact of low, moderate, or high.

Once the calculation is complete, you then apply the highest watermark to the IT system. For instance, let's say we have a Microsoft SQL database that holds PII, including credit card and **Health Insurance Portability and Accountability Act** (**HIPAA**) data. This database is used in a hospital setting and must be available 24 hours a day, 7 days a week, and 365 days a year. Any downtime the database may experience could mean life or death to the patients. Given this scenario, the equation would look like this:

$$SC\ SQL\ Database = \{(confidentiality, high), (integrity, high), (availability, high)\} = High$$

Now, let's say there is another Microsoft SQL database; however, it holds mailing information about its customers. Though it holds personally identifiable information, even if it goes down for a short period of time, it will not hurt the business, so the equation looks like this:

$$SC\ SQL\ Database = \{(confidentiality, moderate), (integrity, moderate), (availability, low)\} = Moderate$$

In this example, the overall security watermark for the database is *moderate*. As confidentiality and integrity of the database are more important than availability, you must apply the highest watermark.

Risk, in a general sense, is subjective. Two organizations that perform the same type of business can have two completely different risk profiles due to their overall risk appetite. To better understand your risk appetite, you will need to discuss this with your senior management.

SP 800-171

If you work for a private or public entity that works on U.S. **Department of Defense (DoD)** contracts, chances are that language around the **Defense Federal Acquisition Regulation Supplement (DFARS)** clause 252.204-7012 has been written into your contract. Many research universities, manufacturing, and other private sector companies receive **Controlled Unclassified Information (CUI)** from the DoD. To better protect sensitive information held by non-federal government agencies, NIST developed SP 800-171 in June 2015. With only 14 control families and about 100 controls in total, SP 800-171 utilizes a subset of the controls found in SP 800-53.

Defined by the National Archives, CUI is information that needs to be protected in both physical and virtual forms. The following categories are types of CUI that should be protected according to SP 800-171:

CUI Categories	
Critical Infrastructure	Defense
Export Control	Financial
Immigration	Intelligence
International Agreements	Law Enforcement
Legal	Natural and Cultural Resources
North Atlantic Treaty Organization	Nuclear
Patent	Privacy
Procurement and Acquisition	Proprietary Business Information
Provisional	Statistical
Tax	Transportation

Table 2.2 – Information that should be protected

NIST Cybersecurity Framework

President Obama signed an executive order in 2014, directing NIST to develop a voluntary risk management and cybersecurity framework that could be used by all industry sectors and any size organization. This framework is outlined in three sections: the Core, Implementation Tiers, and Profile:

- **Core**: The core comprises five functions: identify, protect, detect, respond, and recover. These outline the implementation of cybersecurity best practices, risk management, and technology governance. Each function is further divided into categories and subcategories, with information references (alignment with other security frameworks) that provide direction for the development of enterprise policies, standards, and procedures.

- **Implementation Tiers**: The implementation tiers describe how the organization should approach cybersecurity and the processes used to evaluate risk. Tiers are broken into four separate categories, which increase in risk awareness levels, automation, and participation in threat intelligence. Tiers are defined as follows:

  - **Partial**: Cybersecurity risk management processes are not formalized or reactive to a situation. Priorities related to threat and risk management are not established or well documented. Overall organizational awareness of cybersecurity management and risk is limited due to the minimal knowledge and experience of staff.

  - **Risk-Informed**: There is organizational awareness of cybersecurity risk; however, the approach to managing risk is not formalized. Management has approved appropriate measures to reduce or mitigate risks; however, no formal policies have been established. Prioritization of cybersecurity initiatives is directed by the business requirements and threat landscape of the organization and its peers.

  - **Repeatable**: Organizational policies, standards, and procedures are well documented, approved by management, and regularly updated based on changes in requirements. Employees are informed of such policies, and management ensures they are followed as intended. The organization understands the role it plays in its customer supply chain, takes proactive steps in securing its IT resources, and shares threat intelligence with its community.

  - **Adaptive**: The organization embeds cybersecurity and risk management into the culture of business practices. *Lessons learned* sessions are performed after cybersecurity incidents to better understand the response to the incident and discuss process improvements. Management updates policies, standards, and procedures based on lessons learned session recommendations to improve the response to cybersecurity risks.

- **Profile**: The NIST Cybersecurity Framework Profile is used to describe the organization's current approach to risk management and its desired future state. The Framework Profile is determined by the Core functions, categories, and subcategories, and how these are aligned with the organization's business requirements, risk appetite, and resources. While the current state represents the positives and negatives of the cybersecurity risk program, the future state is used to identify gaps in the risk management program and develop strategies for how the organization will achieve its objectives.

  The future state of the Framework Profile will not only include project objectives but also allows the organization to determine capital and operational expenditures along with establishing timelines. Once identified, management will determine whether it can achieve its goals within the timelines or whether it will require the organization to accept, transfer, or avoid the identified risks altogether. As with any project timeline, the organization can choose whether to have one document that describes the objectives for a future state or develop future state documents that align with multi-year strategies.

NIST has developed hundreds of information technology and information security documents. The documents are all available for free and can be found on their website[1]. If your business is located overseas, a different standard might suit you best, such as ISO/IEC 27001.

ISO/IEC 27001

Located in Geneva, Switzerland, the **International Organization for Standardization (IOS)** has been developing standards since 1947. **ISO/IEC 27001** is an international standard used to secure businesses and evaluate risk. The standard also defines the **Information Security Management System (ISMS)**, which is used to document and store policies, standards, procedures, and system information.

The ISO/IEC 27001 document is a comprehensive approach to identifying and mitigating risk within your infrastructure. As part of the process, it requires the organization to develop or adopt a risk management framework. This risk management framework is then used to evaluate deficiencies within IT systems and implement compensating controls to mitigate those deficiencies. The standard does not require you to use the ISO risk management framework (ISO 31000); however, to comply with the standard, you must have one in practice.

ISO 27001 specifies an ISMS, which is used to ensure that security controls are in place by applying the CIA triad to the data. The CIA triad consists of the following:

- **Confidentiality**: Restricting access to information to only those who need to view it

- **Integrity**: Ensuring the data or information has not been changed or manipulated by an unauthorized third party

- **Availability**: The information is accessible to authorized third parties and follows the required **Service Level Agreement (SLA)** or **Operational Level Agreement (OLA)**

Organizations and individuals can also get certified in ISO 27001. Organizations use the certification process to prove to their customers that a formal security program has been implemented. Individuals can get ISO certified so they have the background and knowledge of implementing and assessing an ISO/IEC 27001 program.

There are plenty of other cybersecurity frameworks developed by IOS – security standards that include cloud computing, network security, application, supply chain, cloud computing, and incident management. They also publish standards for automotive security, privacy guidelines, and risk management.

Cloud Security Alliance

Do you work in a start-up? Do you or your organization have a cloud-first policy? Does your organization use Microsoft Azure, **Amazon Web Services (AWS)**, or **Google Cloud Platform (GCP)**? If you answered yes to any of the previous questions, then you should look at the **Cloud Security Alliance**[2].

Formed in 2008, the Cloud Security Alliance was created to assist organizations in deploying and securing compute, storage, and other services in the cloud. The Cloud Security Alliance has several different offerings to choose from. They have an online security registry called **Security, Trust, Assurance, and Risk (STAR)**. This registry is available to organizations that wish to attest their security controls using the **Cloud Controls Matrix (CCM)**. To align yourself against the CCM, the Cloud Security Alliance also developed the **Consensus Assessments Initiative Questionnaire (CAIQ)**. Consisting of 17 security domains and 197 controls, the CAIQ has become the standard for any *as-a-Service* provider and cloud-first organizations.

Organizations can choose between two different levels of the STAR registry. The first level allows you to perform a self-assessment and upload the results for free. This is a great way to showcase how your organization is meeting the controls. Level 2 is more stringent, requiring the organization to hire an external third-party assessor to perform the assessment. While level 1 allows you to upload the findings and store them for free, level 2 requires the company to pay for their listing. Major cloud service providers, such as Alibaba, AWS, GCP, and Microsoft's Azure, are all members of the Cloud Security Alliance, having submitted their results for review.

Center for Internet Security

In collaboration with security experts from across the globe, the SANS Institute first developed the Top 20 Security Controls in 2008. Continuing this effort, in 2015 the **Center for Internet Security**[3]**(CIS)** took over the Top 20 Security Controls and introduced the concept of **foundational cyber hygiene**. This concept, when implemented properly, could have the potential to significantly reduce an organization's risk of cyber threats. Estimates have it close to around 85% when implementing the first six control families.

At the time of this writing, CIS released version 8 of the Critical Security Controls. Major differences between the latest version and its predecessors are the removal of the foundational cyber hygiene concept and the reduction of controls from 20 to 18. The reason behind the change was to emphasize the focus on the organization's size and maturity of the IT department.

While CIS has not provided quantitative guidance as to the number of full-time employees a company should have on staff, they do provide advice on how to align the IT department to an IG. There are three IGs, and each group builds upon the other. For instance, if you were to assess your company's IT department and decided to align it with IG3, that would mean the company must also implement IG1 and IG2 as well.

CIS now refers to the implementation of the first IG as *essential cyber hygiene*. Foundational cyber hygiene, now essential cyber hygiene, is an important security control as this is a control that every organization should have implemented. Examples of the controls for IG1 include inventorying all known and unknown assets connected to your local area network. The known assets should be placed in a centralized repository called a **Configuration Management Database (CMDB)**. Once the IT department has gone through and inventoried all known assets, what do you do about the unknown assets? That is a question for you and your team to decide. A CMDB that holds all the information for physical assets should also include software installed on those assets. In other words, document, document, document.

In addition to the Critical Security Controls, CIS also produces security benchmarks. These benchmark documents are technical documents that provide guidance on how to harden software and services. CIS has written hundreds of benchmark documents for a good majority of operating systems, network equipment, databases, web browsers, office suites, and many cloud service providers.

DISA STIG

The United States government and military hold onto some of the most sensitive information in the world. To adequately protect this information, the **Defense Information Systems Agency** (**DISA**) created the **Security Technical Implementation Guide** (**STIG**). The STIG is a set of documents used to harden information systems used by the different branches of the military. They are also seen as some of the most stringent security controls out there.

DISA has created hundreds of hardening documents for networking equipment, operating systems, mobile phones, web servers, and databases. In some instances, DISA created two separate documents for the same service. For instance, they have documentation for setting up Apache on both UNIX and Microsoft.

While some frameworks provide paid resources for implementation and assistance, DISA has mailing lists that can be subscribed to. These mailing lists provide community assistance for those who may need a little bit of help implementing new technology or controls. These are all free of charge to the security community.

PCI-DSS

At the time of this writing, the **Payment Card Industry Data Security Standards** (**PCI-DSS**) is in its fourth revision. The PCI-DSS was developed by major credit card companies to standardize a set of security standards to be used by organizations that transmit, store, or process credit card transactions. The PCI-DSS has 12 sections that cover the overall security of the payment card environment.

To become compliant with PCI, you must fill out the appropriate **self-assessment questionnaire** (**SAQ**), which can range from web stores to payment card terminals. There are also four separate levels that an organization will fall into depending on how many credit card transactions are performed within a given year:

- **Level 1**: Process more than 6 million transactions per year
- **Level 2**: Process between 1 million to 6 million transactions per year
- **Level 3**: Process between 20,000 to 1 million transactions per year
- **Level 4**: Process less than 20,000 transactions per year

These levels are important as PCI dictates whether an organization should complete an annual SAQ on its own or whether it requires the company to hire out the review of security controls. Some rules require that a company has a third-party audit if the organization has experienced a data breach within the last 12 months.

All too often, I have seen organizations implement security controls and never revisit them. Cybersecurity is a mindset, not something that you set and forget about. The next section will discuss this in greater detail.

Examining security as a checkbox

Many organizations and security professionals make the mistake of implementing security controls and calling it quits. Security is not, and never has been, something that you implement and forget about until the next assessment. Take for example PCI's SAQ. If you take or process credit card information, you must fill out and submit an SAQ to your processor on an annual basis. This is used to show your level of compliance with the regulation.

For many years, organizations were passing their PCI assessments with flying colors. However, in just a few short months after getting their PCI certification, a breach occurred. How can this be? The organization was PCI certified! In many cases, it's because the business saw the regulation as a checkbox, a process, or a set of hoops to jump through to obtain the certification. Once the auditors left, security became a second thought to them.

As a security professional, it is your responsibility to ensure the organization maintains a baseline of security. The baseline does not increase just because you have an upcoming audit and then decrease afterward. The baseline is just that – a baseline. That baseline should be maintained throughout the year, not just on special occasions. Therefore, many regulatory bodies require periodic testing and reporting to ensure the business is maintaining that baseline.

For instance, PCI requires the submission of vulnerability scans quarterly to maintain certification. Other certification bodies observe businesses, their security controls, and their processes over a period of time (sometimes months) rather than just a point in time. These are all measures to ensure that the organization takes security seriously and it is not something of an afterthought.

There are plenty of ways to ensure compliance. Automation software such as Puppet, Chef, and Terraform can be deployed to ensure configurations are in place for IT systems. Vulnerability management software is used to ensure systems are free from defects or exploits. Benchmarks can be used to harden IT systems. Risk management processes can be developed to review system risks and apply security controls where necessary.

Cybersecurity is more than checkbox security; it is a mindset. As a cybersecurity professional, especially starting out, it may feel like you are constantly drinking from the firehose. In most instances, you are, and that feeling may never go away, but that is the excitement of working in an ever-changing landscape. Frameworks are the same: always changing to stay up to date with the latest threats.

Security and continual improvement should go hand in hand when discussing cybersecurity. Not only should you implement the necessary controls, but you should evaluate the controls at a given interval to ensure they are working as intended.

Understanding continual improvement

Many security standards and regulatory bodies encourage, and sometimes require, you to continue to grow your security program over time. Frameworks such as the NIST Cybersecurity Framework introduced Implementation Tiers, used to show how an organization is improving on processes. The Software Engineering Institute at Carnegie Mellon University developed the **Capability Maturity Model Integration** (**CMMI**), a maturity index used to help organizations evaluate themselves. The levels of CMMI are as follows:

1. **Initial**: Processes are not defined or documented

2. **Managed**: Processes are defined, some are documented, but most are reactive in nature

3. **Defined**: Processes are defined, documented, and signed off; the organization is somewhat proactive rather than reactive

4. **Quantitatively managed**: The organization is proactive and uses current data to implement predictive analytics to fend off adversaries

5. **Optimized**: The organization or cybersecurity team is heavily focused on process improvement and automation

To put this into practice, we can use the **Deming Cycle**. William Deming, a quality control engineer in the 1950s, developed the concept of **Plan, Do, Check, and Act** (**PDCA**). Originally created to improve business processes, we can use this same concept in cybersecurity:

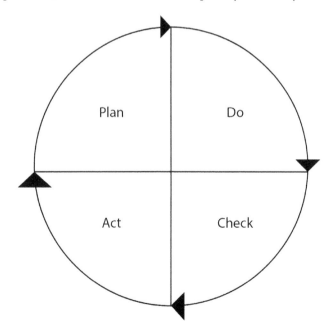

Figure 2.1 – The Deming Cycle

To put PDCA into practice, we must first understand how it can be used:

- **Plan**: During the planning stage, you work with various teams to ensure you have all the business requirements to implement a new IT system. During this phase, you must also plan for the various types of security controls that will be implemented in the system. Later on, in *Chapter 6*, we will discuss the Systems Security Plan, where you document the IT system, its security controls, points of contact, and so on.

- **Do**: After the planning phase has been completed, it is time to implement the IT system. The IT or security team is now ready to install the system and implement the controls laid out during the planning phase.

- **Check**: Once the new IT system has been implemented, we check it to ensure that it was implemented properly, and review the planning documents to ensure that the system was implemented as designed. It is also your responsibility to review the security controls implemented on the new IT system, making sure that the controls are working as intended. This can be performed through automated or manual means of checking. There is also no timetable for when to perform the checks; it is based upon the risk established by the organization. However, some regulatory requirements, such as PCI, require these to be checked periodically to maintain compliance.

- **Act**: The last phase of the Deming Cycle is *act*. This is when you have reviewed all the controls of a system and need improvement. For instance, you have implemented a new web server. The web server is running as intended with all of the discussed controls implemented. During the *check* phase, however, you noticed that information regarding vulnerabilities in the system is not being updated as frequently as the security team would like. To compensate for this, the security team must implement a change in the reporting frequency to obtain better scan results.

To implement this change, we must start at the beginning of the cycle and run it through its phases once again. The frequency must be updated on the system and in the Systems Security Plan. Once documented, we need to implement the changes and test them to ensure they are working as intended.

While the Deming Cycle and CMMI are not discussed in the previously mentioned security frameworks, they should, however, be part of your process improvement strategy. By continuously improving your processes, your IT and security team will make fewer mistakes. It will also reduce the amount of rework that may have to be done to secure the system as per the documentation. By utilizing these improvement processes, you can be assured that you and your security program are on the right track.

PDCA can be used with any framework you choose. The point here is that you should work to continuously improve your program, and PDCA is one way to do that. When selecting the framework that is right for your company, remember not to implement security controls just to fill a checkbox.

Selecting the right framework

There is no right or wrong framework to choose; however, you should select a framework that closely aligns with the business function. If you are an international company, then the ISO 27000 series would be a great choice. If you are U.S.-based, NIST might be the best way to go. Do you hold on to U.S. federal government contracts? Then you should research CMMC in addition to DFARS and NIST 800-171 to see whether that will provide the necessary security requirements.

You may need to combine multiple frameworks into one for your organization. Maybe part of your network receives credit card transaction data. In addition to PCI, part of your network holds on to ePHI data, and you have adopted the NIST Cybersecurity Framework. If the systems and network are not set up properly, then your entire infrastructure is in scope. Covered more in *Chapter 3*, you will learn that scoping your environment is not only necessary but will save you time and money when an assessment is performed. Once the environment is scoped properly, you can then apply the framework and controls to a subset of systems rather than the entire infrastructure.

Frameworks are made to provide guidance on what controls to implement and how they should be configured. You could make your own framework, combining controls from multiple different frameworks. By doing this, you can pick and choose which controls to implement. If you have a small IT or security team, then it might make sense to pick a subset of controls and put those in place, much like the implementation groups discussed in the *Center for Internet Security* section.

A word of caution when picking and choosing which controls you choose to implement when making your own framework: the security manager must understand which controls are not being adopted and, in doing so, understand that you are inheriting risk. A thing to note as a security professional is that you act as an advisor to the business.

While acting as an advisor to the business, you still need to make decisions that will align them appropriately with the security objectives. Using frameworks is one of the best ways of getting there.

The framework used in this book

Though I have highlighted a few different cybersecurity frameworks, we will focus our efforts on the NIST **Cybersecurity Framework** (**CSF**) and the CIS CSC. The NIST Cybersecurity Framework is a great framework to use, but it is focused more on administrative controls – those being policies, standards, and procedures. The CIS CSC focuses heavily on technical controls. Both frameworks are great to get an organization of any size and industry aligned with cybersecurity best practices.

Another reason is that they are free to use. Well, not really; your tax dollars paid for them. However, they are available to download without having to shell out additional dollars to spend. You will have to fill out a form to download the CIS CSC but do not worry, the people at CIS are great (investigate becoming a member of CIS; there are so many benefits you will get outside of just the benchmarks). As you will learn in *Chapter 3*, performing assessments is important to maintaining your cybersecurity program. These will highlight deficiencies in the program, and you can use the documents to overcome them. How can you do this? Stay tuned!

Summary

There are hundreds of cybersecurity frameworks available to choose from; how do you know you are choosing the right one? Talk to stakeholders in the business to understand what the business does, who its customer base is, and where it does business. Also, as established in this chapter, understand how the business approaches cyber risk and how that impacts the enterprise.

Once you understand the business, choose a framework, and begin working on implementing controls. Many frameworks are focused on administrative controls, while others are focused on technical controls. There is nothing that says you cannot combine multiple frameworks to compensate between control families.

Many frameworks have matrices that show how they align with other frameworks. For instance, the NIST Cybersecurity Framework and CIS both have mappings to ISO and SP 800-53. This means that you can begin implementing one framework and if you decide to migrate or adopt another framework, you are not starting over from scratch. With that knowledge, in the next chapter, we will focus on how to use your selected framework to perform a cybersecurity assessment.

References

1. The National Institute for Standards and Technology Special Publications: `https://csrc.nist.gov/publications`

2. Cloud Security Alliance: `https://cloudsecurityalliance.org/`

3. Center for Internet Security: `https://cisecurity.org`

3

Cybersecurity Strategic Planning through the Assessment Process

You have been placed into your dream job, built some internal relationships, and picked a cybersecurity framework; now it is time to put them to use. First, you will need to perform an assessment to better understand the current state of your organization's cybersecurity program. This can be accomplished in many different ways; however, one of the best ways to obtain this information is through performing a cybersecurity assessment. Once you have completed the assessment and concluded the current state, you need to understand the gaps between the current state and where you want to take your program.

In this chapter, we will learn how to perform such an assessment for an organization's cybersecurity program. From there, we will develop our cybersecurity strategy, which is essentially a plan to get us to the future state of where we want to take our cybersecurity program or what we want to achieve. Lastly, we will develop SMART goals to define how we will resolve those objectives and when. This will all help resolve the issues that we uncovered and how we intend to remediate them.

In a nutshell, we'll be discussing the following topics:

- Developing your cybersecurity strategy
- Who should perform an assessment
- Preparing for the assessment
- Performing the assessment
- Wrapping up the assessment
- Understanding the current and future state of your program
- The exit interview

Developing your cybersecurity strategy

You must begin developing your strategic plan, but how do you get there? You cannot plan your end goal unless you understand where you are starting from. The only way to better understand where to begin is by looking at historical documents – past assessments such as penetration tests, audits, and vulnerability scans. This is documentation that you must review to get a sense of what the current organizational cybersecurity risk is. If these documents are not available, or are more than 6–12 months old, then it is time to get a fresh perspective. You should also review the organization's architectural drawings, policies, standards, and procedure documents. These will help set a foundation for the assessment and provide the assessor with a brief overview of the environment.

As the head of security, you must understand what those risks are and how to overcome them. This can be performed through a holistic assessment of organizational security. To do this, align the assessment to a given standard. As mentioned in the last chapter, we will use examples in this book aligned with the NIST Cybersecurity Framework and the Center for Internet Security. However, you could use ISO 27001, the Cloud Security Alliance **Consensus Assessment Initiative Questionnaire** (**CAIQ**), or any other framework that aligns with your business goals and objectives.

Who should perform the assessment?

Cybersecurity assessments take time – time to develop, time to prepare, time to perform, and time to wrap up. *Who should perform the assessment* is an important question to consider. Assessments can be performed by yourself (first party), your suppliers or customers (second party), or by an outside firm (third-party). If you do not have the time to put into an assessment, you will not get the results you need to develop your plan. Not all third-party auditors will perform the assessment with the same vigor as you. You must ensure that the information gathered during the assessment is correct to make actionable decisions for the security program. Incorrect information could lead to a false sense of security.

A third party assessment is performed by a trusted external organization. Third party assessments are beneficial for several different reasons. First, they allow for a non-biased review of your environment. This takes the ambiguity out of whether the cybersecurity assessment was skewed in any manner. The potential for *sugar coating* or *being partial* is also removed if it is you doing the assessment. Most certification bodies, such as the International Organization for Standardization, require an independent third-party assessment to be performed. Others, such as the payment card industry, allow for self-attestation only up to a certain amount of credit card transactions. Those who fall under a tier-1 credit card processor are required to have an independent body perform these assessments.

First and second party assessments can be as stringent but not looked upon as favorably as independent third party assessments. Second party assessments are also performed by external parties; however, these are performed by organizations that already have an established interest in your company. Let us say that you retain sensitive information for a customer. That customer has the right to assess your cybersecurity practices and send you a questionnaire to fill out. It is then up to the customer to work with you on implementing any controls deemed necessary in the environment.

First-party assessments, or those performed by you or an internal audit team, can be the most cost-effective. By performing an assessment yourself, you have the ability to answer the questions that are important to you and your organization. However, as the person performing the assessment, you must not be biased when answering the questions. If you do not take an objective look when performing the assessment, this can give you a false sense of security.

Too often, I have seen organizations fill out first-, second-, and third-party assessments questionnaires, massaging the questions to make them look favorable. This also gives a false sense of security and could come back to haunt you if a cyber-incident ever occurred. This is also not ethical, and I suggest that you do not do this. Being truthful is always the best answer, even if it is not a positive one.

When performing a first-party assessment, consider the following phases for the engagement:

1. An engagement letter
2. Project initiation
3. Performing the assessment
4. Gathering findings
5. The exit interview
6. Assessment closure

Performing an assessment yourself can sometimes feel like more of an art than science. In the next section, we will discuss how to perform the assessment on your own. Using this approach, you can choose which cybersecurity framework you want to use – even if it is one that you made yourself. This can be done on your timetable and can save money.

Preparing for the assessment

When performing an assessment yourself or by an internal audit team, you should be aware that not everything can go as planned. While it does not happen often, performing manual or automated scans can be detrimental to an IT resource. For instance, you could perform an NMAP or CIS-CAT scan for how a system is configured. If the IT resource is a few years old, it may not keep up with the demand of processing the scan and fail. In the next section, we will discuss why having an engagement letter is important if you decide to perform any scans against IT resources in the organization.

Drafting an engagement letter

The engagement letter is simply a letter from an organization's CEO or someone in authority that can sign off on the engagement. While this is a small step in the overall grand scheme of things, it is highly important. The engagement letter should state the scope of the assessment, its purpose, who is performing the assessment, the key stakeholders, the approximate duration of the assessment, and an overview of the standards being used.

A test plan can also be included with your engagement letter. The test plan will cover how you intend to perform the assessment. Whether it is interviewing personnel, performing vulnerability scans (manual or automated scans to review configurations of systems), an over-the-shoulder review, or through **open source intelligence (OSINT)**[1], it needs to be spelled out in the document so that there is a clear understanding of what will and what will not be performed during the engagement. Think of this as your **statement of work (SOW)** between yourself, your team, and the business.

This is your *get-out-of-jail-free* card. While there is no reason to think that any scans being performed would affect an IT resource, there is always that chance. The engagement letter should also state if and when scans will be performed in the environment. That way, if required, you can schedule scans outside business hours so that you do not impact production. If you do take a system down, you will have a letter stating that management is allowing you to do this. The next phase of the project is to begin running scans and information collection.

Project initiation and information gathering

The project can be initiated through a preliminary questionnaire. This questionnaire can provide valuable information to you, as the one performing the assessment, if you are not fully up to speed on the environment. As the new head of security, you may not fully understand the environment yet and want to ask additional questions that are not found in a cybersecurity standard. This can also be the time to perform additional security scans of the environment, allowing you to gain valuable information on the overall risk posture.

There are plenty of free OSINT tools available to use to understand the risks (and there will be more on risk in *Chapter 6*) in the environment. With Recon-NG, you can perform OSINT scans of your environment. This allows you to gather information about your organization that is freely available on the internet. **Dnsenum**[2], another open source tool, will allow you to perform DNS lookups, allowing you to gather information on what systems and possible services are being provided on the internet. You could also perform port and vulnerability scans using **Network Mapper (NMAP)**[3], allowing you to understand the various types of weaknesses an IT resource may have.

This is also the time to better understand what the business needs are. As the person performing the assessment, you will need to figure out what is important to the business. This will ensure that the way questions are asked is relevant to the business. If ransomware is of concern to the organization, you should ask how backups and disaster recovery are being performed, when they are being performed, and whether successful restores have occurred. These follow-up questions are needed to better understand the process and are directly related to the backup process.

It is also time to gather any documentation about the environment. This includes architectural drawings, previous assessments, vulnerability scans, penetration tests, and any and all documentation that will be helpful for the assessment. The key here is to continue to understand the environment prior to the engagement. However, there should be a cut-off time for the documentation. Anything older than 6 months to a year is should be discarded.

Performing the assessment

Now, it is time to get your hands dirty! The NIST Cybersecurity Framework is ideal for administrative tasks, while the Center for Internet Security is great for technical controls. While the technical side of performing assessments may be fun and exciting, do not forget the administrative side of it too. Each framework comes with its own set of controls but complements others nicely.

As mentioned in the previous chapter, the NIST CSF has five functions, each with its own categories and subcategories. They are as follows:

1. **Identify**: The identify function is geared toward the development and understanding of cybersecurity risks to IT resources. This function is used to better understand the business and its functions, and sets the foundation for how to begin planning your security meeting.

2. **Protect**: The protect function is used to implement safeguards for IT resources. It can also be used to isolate or contain a cybersecurity incident from spreading. The categories for this function include cybersecurity awareness and training, data protection, and safeguards for maintenance.

3. **Detect**: How would you know whether the safeguards put in place are working? There must be a way to detect anomalies, threats, and vulnerabilities within the environment. Cyber threats and incidents occur regularly. How would you know whether you are being attacked if you did not have a mechanism in place to detect it?

4. **Respond**: Once a cybersecurity incident has been detected, how will you respond to it? What safeguards are in place to isolate or mitigate a threat within your environment? The respond function will help you identify weaknesses in your incident response plan and show how to resolve them.

5. **Recover**: We have identified the environment, placed adequate protections in place, developed tactics to detect threats, and have an incident response plan in place. Lastly, we need the ability to recover from an incident. The recover function is used to help create a business continuity and disaster recovery plan.

Here is the full list of categories and subcategories that fall under each function:

Function ID	Function	Category ID	Category
ID	Identify	ID.AM	Asset management
		ID.BE	Business environment
		ID.GV	Governance
		ID.RA	Risk assessment
		ID.RM	Risk management strategy
PR	Protect	PR.AC	Access control
		PR.AT	Awareness and training
		PR.DS	Data security
		PR.IP	Information protection processes and procedures
		PR.MA	Maintenance
		PR.PT	Protective technology
DE	Detect	DE.AE	Anomalies and events
		DE.CM	Security continuous monitoring
		DE.DP	Detection processes
RS	Respond	RS.RP	Response planning
		RS.CO	Communications
		RS.AN	Analysis
		RS.MI	Mitigation
		RS.IM	Improvements
RC	Recover	RC.RP	Recovery planning
		RC.IM	Improvements
		RC.CO	Communications

Table 3.1 – NIST CSF functions, categories, and subcategories

Each category has a subcategory with an associated information reference. This reference assists in determining which CSF controls line up with controls of other cybersecurity frameworks. For instance, subcategory ID.AM-1 has a control objective of "*physical devices and systems within the organization are inventoried.*" This control aligns with the Center for Internet Security **Critical Security Control (CSC)** 1 (which also pertains to inventorying IT systems). ID.AM-2 has a control objective of "*software platforms and applications within the organization are inventoried.*" This aligns with CSC 2. When performing an assessment, you must take into account whether or not the control is in place and whether there is a policy document that backs this up. Auditors want to ensure that not only does the organization have the control in place but also that employees and IT resources are following the documentation.

Let's say that you are performing an upcoming assessment. The documentation is all ready to be presented to the auditors, and systems administrators are on standby to assist. Within the documentation, there is a standard that covers antivirus software. It states the type of software, the frequency of virus definition updates, and how often it performs a system scan. Let's say an auditor asked to review a sample of systems to ensure the antivirus software is configured to the standard. Upon review of one system, it was determined that a systems administrator forgot to install the software on one system, and on another, the antivirus software was not updating as frequently. Given this scenario, you would receive two findings – one for missing software and another for not updating per the documentation.

There are several different ways to conduct an assessment, which consists of sending out a questionnaire or conducting it in person (a sample questionnaire can be found on GitHub at `https://github.com/PacktPublishing/Executive-s-Cybersecurity-Program-Handbook`). Conducting an in-person assessment where you interview those within the organization should occur at given intervals, whether that is annual or biannual. This is to ensure that questions are answered objectively and biases are removed. It is also important to pay attention to what is not being said during the interview. Body language, nonverbal cues, and hesitation from the person being interviewed should be clues for follow-up questions. Overhearing sidebar conversations will provide additional insight into how you answer the questions. This is not the time to interrogate the person, but there are clues to pick up on during the interviewing process that can often lead to a thorough answer.

When performing an assessment, you must ensure that all the control objectives have been reviewed and answered. For instance, when collecting information pertaining to ID.AM-1, ask the question in a way that makes sense to the interviewee, such as, "How does the organization keep track of physical IT resources?" Engage with the interviewee and ask follow-up questions, such as, "Who is responsible for putting systems into a centralized tracking system?" As previously mentioned, there should also be a policy or standards document that discusses this in detail. This will help you when scoring your assessment, but there will be more on that in the next section.

Wrapping up the assessment

Once you have wrapped up performing the assessment, it is time to evaluate the results, score the assessment, and prepare for the *report out* or *exit interview*. When evaluating the results, how did you do? Did you get all your questions answered? Were there any noticeable gaps during the assessment process?

When starting your cybersecurity program, you may not have all the answers immediately or even at all. Some organizations excel in certain areas but lack in others, and that is okay. Remember, you have to crawl before you can walk, and boiling the ocean is not a way to get there. Where a program is lacking is what an assessment is supposed to answer.

That being said, there are two ways you should evaluate the results of the assessments – administrative and technical. Technical controls are like the ones you see on television – a person, in a dark room, hammering away at the keyboard writing lines of code on a black screen with green letters. While this sounds more like a hacker story, it does portray what someone could be doing at a technical level. You will never see a movie of someone just writing IT policies all day. However, the importance of administrative policy is critical to your overall cybersecurity program.

Let's look at the review methods in a bit more detail.

Administrative review of policy documents using the NIST CSF

Assessing the administrative controls of an organization comes down to policy documents. Their importance relates to documentation – can your organization back up what it is doing policy-wise? Can the organization perform the same in reverse by providing proof that what they have as a policy is being performed on IT resources?

Scoring the assessment from an administrative perspective can be subjective as well; however, there is a method that I use. By taking the CSF's implementation tiers and assigning them a score, you can determine where on the maturity index you are. This is just one step in the process of developing your current state of cybersecurity. That being said, the assessment and scores are a point-in-time assessment and should be treated as such. Performing quarterly spot checks against the NIST CSF should be part of your overall improvement process.

The following are suggestions for how to score yourself. We will take the implementation tiers and assign a rating to them. Each tier builds upon itself to show how a program is maturing:

- **Tier 0: None – score of 0%**: The organization does not have any controls in place or does not feel that the control objective pertains to them.

- **Tier 1: Partial – score of 25%**: Risk management practices are not formalized, and processes are ad hoc or reactive. There are no policies or standards written for this control.

- **Tier 2: Risk informed – score of 50%**: Management has approved risk management practices; however, they have not been approved company-wide. Processes are still ad hoc and reactionary; however, organizational risk is identified through a manual or automated process.

- **Tier 3: Repeatable – score of 75%**: Risk management practices are approved organizational-wide and documented in policy. Risk-informed decisions are made about cybersecurity practices and are updated on a regular basis.

- **Tier 4: Adaptive – score of 100%**: Cybersecurity practices are continuously informed and updated, based upon risk management. Activities are run after cybersecurity incidents to learn lessons and inform and improve upon the security program.

Score each subcategory based on the tiered scoring model. To determine the scores for each category and function, calculate the average score already determined from the subcategories. Here is an example:

Function	Category	Subcategory	Score
RS	RS.MI	RS.MI-1	25%
		RS.MI-2	75%
		RS.MI-3	50%
Average subcategory score			50%

Table 3.2 – NIST CSF response mitigation category

The average for RS.MI is 50%. When looking at the scores, it is easy to tell where exactly you currently are on the maturity index. Score all the categories and subcategories and find the averages. This will help you determine the current state and maturity of your cybersecurity program. This will also assist in seeing how you score against the NIST CSF tiers.

A technical review using the CIS controls

Technical controls are the opposite of administrative controls. If you have a policy that states that virus definitions are updated every 6 hours, is that actually being done on the server? Do you have a **configuration management database** (**CMDB**) in use that is being actively maintained as per the standard?

When performing this technical assessment using the CIS controls, you first need to determine the **implementation group** (**IG**). The IG is based on the size of an organization's IT staff. If you work for a small organization that may only have 1 or 2 IT people, you will fall under IG1 and so on. Each IG has a set of controls that are recommended to be in place within the organization.

With the CSC, we are not so worried about people or the technology being used; we are more concerned about the process. In section 1 of the CSC, *Inventory and Control of Enterprise Assets*, IG1 requires that you implement an asset inventory database (CMDB) to track all authorized and unauthorized IT resources. IG2 of section 1 requires that you not only implement the controls of IG1 but also implement DHCP logging and an active discovery tool. As you can see, the CSC does not mandate that you use a MySQL or Microsoft SQL database to store that information. You are free to choose the technology that suits your organization the best.

It is time to begin interviews and ask questions in relation to the IG and CSC. These can be true/false types of questions. However, you should understand the environment and how systems were configured and why so that you can make risk-informed decisions. These decisions will dictate your future state and an overall strategy for how to get there. You could answer true to a question of whether you have a centralized asset inventory database, but does that truly provide the answer you are looking for? How would you know if all IT resources are in the database? Is there a certain threshold where the business decides they do not want to track a piece of equipment? This could be an area for improvement, which you would not know unless you asked the right questions.

If you purchased the Center for Internet Security SecureSuite Membership and have access to CIS WorkBench, you can download their CIS-CAT Pro scanner. The CIS-CAT Pro scanner will allow you to perform automated scans against IT resources and provide a detailed report of the findings. From there, you can make risk-based decisions on how you would like to move forward in remediation.

In addition to the CIS-CAT Pro scanner, CIS also offers the CIS CSAT tool. The CSAT tool is provided free of charge and can be used as a means for answering questions and tracking your progress. It allows one organization to have multiple employees using the tool and assign tasks based on the answers. This tool is free of charge and can be leveraged without a membership.

Understanding the current and future state of your program

By performing an assessment and compiling the results, you can get a good understanding of the current state of your cybersecurity program. The current state is a point-in-time evaluation to remediate any findings and determine the next steps. The future state is where you as the head of security develop the strategic plan.

This future state is intended to be your plan for the next 2–5 years. What controls do you have in place currently? Where are you lacking? What controls seem to need some extra TLC? These are just a few of the questions that need to be answered to help develop your future state. You will also need to consider capital and operational expenditures.

Using the scoring model mentioned earlier, you can determine where you excel and where controls could use a little extra help. You will need to decide what is considered *low-hanging fruit*, what will be a major project, and what is important to the business. Low-hanging fruit can be easy things to resolve, such as making DNS changes to help mitigate spam and phishing emails, adjusting policies on your firewalls, configuring your identity and access management tools to alert you about suspicious logins, or creating risk-based profiles for your high-risk users, such as those in the C suite or your finance department.

A major project can take a considerable amount of time to plan and coordinate with IT and the business. It could also take time to fund it and get it placed on a project roadmap. These factors can also have a substantial impact on the business and its employees. Projects such as multifactor authentication can affect everyone in the business and must be planned out. Working with the business, you can determine what is important to it and understand its concerns and risks to plan accordingly.

To assist you in creating a scoring sheet for the current and future states of your program, an Excel document has been created for you and can be found on GitHub at `https://github.com/PacktPublishing/Executive-s-Cybersecurity-Program-Handbook`.

Developing goals

As we plan for our future state, we first must compile our current state scores and develop a gap analysis. This gap analysis will be used as our strategic planning to take the security program to the next level. Executives and the board will want dates for when these objectives will be met. To do this, we will create SMART goals.

Use SMART goals when planning your future state. By using SMART goals, you can clearly define and plan the types of cybersecurity controls that must be implemented. You can set timelines for when a particular task is scheduled to be completed. Measure success criteria to ensure a project is on track with objectives.

SMART goals are as follows:

- **S: Specific**:

  - What needs to be implemented as part of your future state?

  - Who will be responsible for its implementation?

  - How would you go about achieving your goal?

- **M: Measurable**:

  - What metrics will be used to measure the project's success?

- **A: Achievable**:

  - Can you achieve your specified goal?

  - Is it attainable in the time and money allotted for the goal?

- **R: Relevant**:

  - Is the goal relevant to the overall cybersecurity program?

  - Will the goal meet the objectives set forth by you and the business?

- **T: Time-bound**:

  - When should this goal be achieved?

  - Can it be achieved by its required deadline?

There are a few different ways to plan for your future state as well, those being tactical and strategic. Tactical planning is short-term plans that you anticipate completing within 1–12 months. These plans can be considered as your low-hanging fruit or short-term projects that will make an impact on your cybersecurity program. Strategic goals are those that are long-term and usually last 1–5 years. These plans should all be scored and updated regularly to show the progress that you are making. Once you have determined what your future state will look like, it is time to move on to the exit interview.

The exit interview

Once you have completed the interview process, determined the current and future states of the cybersecurity program, and collected all the documentation from the preparation phase, it is time to present your findings. This is where you, as the lead assessor, discuss how the assessment went, your findings, and how you plan to remediate those findings. Be sure to provide capital and operational expenditures, too, as this will be important when discussing the next steps with your stakeholders. When presenting your findings, it would be appropriate to not only have a written document prepared but also present your findings to your stakeholders.

The audience of the exit interview should contain those in the organization with the authority to make decisions. This means that you will have to present your findings to other directors/C suite members. You will have to understand the audience and how you want to present your findings.

Take an objective approach when presenting your findings without placing blame or showing biases. As the head of security, you must communicate your findings to those who may not understand what you are talking about. This is where building relationships comes in so that you know how to communicate with those around you. It is your responsibility to evangelize your cybersecurity program and present to people how you plan to tackle your findings.

Summary

As you can tell from this chapter, determining the current and future state of your cybersecurity program takes time. Using a common standard such as the NIST CSF or the CIS controls is just one way of getting you there. Performing the risk assessment described in this chapter for your organization not only fulfills your requirement for understanding the risk landscape and the overall security posture but also helps determine where you would like to be. This phase of your tenure as the head of security does take a while, but you will get there – just have faith in the process.

When performing the assessment, make sure you are also scoring yourself accordingly. It can be easy to massage the answer to a question to make it look like you have adequate protections in place when, in fact, you may not. To move yourself up from tier 2 to tier 3, the process must be documented and signed off by an executive at the company.

Therefore, the next chapter will discuss how to create policies, standards, and procedures and the process by which they are reviewed and approved.

References

1. Open source intelligence: https://en.wikipedia.org/wiki/Open-source_intelligence
2. DNS enumeration: https://github.com/fwaeytens/dnsenum
3. Network mapper: https://nmap.org/

Part 2 – Administrative Cybersecurity Controls

Would you know how to perform a job function if a senior engineer were to win the lottery? Do your end users understand what *is* and *is not* acceptable use for their company owned device? Are other departments in the organization purchasing network equipment without your knowledge? Administrative controls are used to combat these issues (and many others) that the IT and cybersecurity teams face on a daily basis.

Though it may not be as popular as hacking or writing code, developing administrative policies is a necessity in IT and cybersecurity. Administrative controls are used to establish governance in the workplace and the tools used. All too often though, it can be hard to find exactly what you are looking for.

Policy documents typically range from 20 – 30 pages in length. This makes searching for exactly what you are looking for extremely difficult. Policies, standards, and procedures should be separate documents and specific to the subject. This not only makes it easier to look for what you need, but also allows you to release certain documents to third parties or the public without fear of giving away too much information.

This second part of the book will focus on how to create a policy document and establish a framework for how policies, standards, and procedures should be created. It will also discuss the key phases of creating the document and its review cycle.

4

Establishing Governance through Policy

Writing policies is hard; writing good policies is even harder.

In this chapter, we will introduce governance through policy. Documents outlining **policies, standards, and procedures** (**PSPs**) are an integral part of establishing governance in your organization. However, writing policy documents is just a small part of the overall policy document life cycle.

Once a policy document has been drafted, it must be presented in front of a policy steering committee and then signed off by someone in the C-suite. As this process unfolds, your overall maturity in cybersecurity will improve. This will move the organization upward in the tiers when it is scored against the NIST **Cybersecurity Framework** (**CSF**).

I had a CISO tell me once, "Never be afraid of an audit. It is there to help, not hurt." While this is a very true statement, those on the receiving end may not think that. In reality, though, an audit is meant for one thing: to help an organization improve. Aside from technical configurations of an **information technology** (**IT**) resource, auditors will want to see policy documents meant to assist with the overall IT governance.

In the previous chapter, we learned what it would take to assess an organization. This chapter will highlight the administrative controls needed. Administrative controls, as you will see, are just as important as technical and physical controls. In a nutshell, the following topics will be discussed:

- The importance of governance and policy documents
- Exploring PSPs
- Policy workflow
- Aligning policy objectives

The importance of governance

Governance, or to govern, sets the direction for how the company will operate. Governance is used to establish the foundation for how employees work and interact with others. It is also used as oversight for how something is purchased, implemented, and maintained. Without governance, how would your employees know what conduct is appropriate in the company? How would an employee know how to carry out a particular task if it is not documented? Who is responsible for setting this direction of the company? What is an acceptable use of company resources?

Governance should come from the top down or from those that are empowered to make decisions on behalf of the company. This usually means that governance must come from the **chief executive officer** (**CEO**), or their proxy, and the board of directors. Written PSPs are part of the overall governance of the company.

Effective governance should also be enterprise-wide. Organizations should avoid establishing a common set of controls for technology in one group and introduce a whole different set of controls in another group. This means that one group should not be expected to enter a work order (an incident log) when they need assistance, while the other group is allowed to *shoulder tap* and ask for assistance without entering a log.

Through governance, an **enterprise architecture** (**EA**) team should be created. The EA has many different facets for how to implement governance throughout the organization. EA team members are required to draft policy documents for the IT department and establish standards for the technology being implemented. These must all be implemented based on the business objectives set out by the executive team and the board.

For example, your organization will have firewalls that are constantly being refreshed. Based on the business objectives, you may be required to have an intrusion detection and prevention system, along with data loss prevention. While reviewing the various types of firewall manufacturers, you and your team will settle on a vendor that can do all of these. The chosen technology shall then become the standard vendor to be used throughout the organization. To establish this as a standard, a policy document should also be created. This policy document will go into detail about the types of features the technology must have and the vendor to be used.

Establishing governance within an organization, no matter the size, can be difficult. If governance is not set early, then IT and business units can go off and do their own thing. This lack of governance can lead to supportability issues as architects and engineers have to learn multiple technologies to support business objectives. Have you ever tried to be a master at more than one thing? Maybe a few?

Governance can come from external sources as well. In *Chapter 2*, we discussed several different cybersecurity standards. The **payment card industry** (**PCI**) requires organizations to follow its regulations if it wants to perform credit card transactions. This too is setting governance for an organization. You must conduct business according to their guidelines or face the consequences of not being able to process credit card transactions.

The importance of policy documents

Policy documents help with establishing governance throughout the organization. Without written documentation for how to perform a particular job function, how would the employee know what to do? An engineer could know how to do it through their years of experience, but is it done per the business objectives? And yes, you can verbally state how to perform a particular job function, but how would you know whether the verbal statement is correct? Written documentation provides several benefits for employees.

First, the policy document (a policy, standard, or procedure) should be presented in front of a steering committee. This steering committee is empowered to initially sign off on these written documents, stating that this is how the department or organization will perform a particular function. Second, after the committee has signed off on the policy document, it must be signed off by a member of the C-suite. Third, the policy document must be *evangelized*, in that it must be announced that a policy, standard, or procedure has been written for employees to review. Once the policy document has been approved, the employees should read, understand, and sign off on the PSP. This will provide *teeth*, or documentation, if or when an employee does not follow the policy.

Policy documents are typically split up into several sections, which can include the following:

- **Purpose**: What is the purpose of the policy document? Why was it written? What does it support or establish?

- **Owner**: This should be the job title or the department responsible for drafting the policy document. When selecting an owner for a document, it should not be tied to a person's name. If that individual were to leave the company, then you would need to update all the documents that have that person's name tied to them.

- **Scope**: What is in the scope of the policy document? Could it include third-party vendors or contractors? This is where you state what and who the policy document affects.

- **Policy statement**: When writing a policy statement, it must be clear, concise, and to the point. The policy statement is where you set out how a technology will be implemented (through procedures), how a technology should or should not be configured (through standards), or your high-level topics found in policies.

- **Information references**: The last step in the policy document, and one that is often overlooked, involves specifying how this policy document lines up with your standards. Does this document align with NIST CSF or your chosen framework?

To assist with establishing governance and writing policy documents, the next section will lay out how to successfully differentiate between PCPs.

Exploring PSPs

This section will provide an overview of what PSPs are. All too often, I have seen policy documents that are 20 to 30 pages long. This is too much for a single document. A policy should not state what a technology is, nor should it state how it is implemented. There should be a separation between PCPs. This will make it easier for you, your team, and the organization to locate what you are looking for in a time of crisis or just normal day-to-day activities.

Policies

Policies are overarching documents that provide an overview of the control objective. They are high-level documents that establish governance with the intent of applying administrative, technical, and physical controls. Policies are high-level, allowing anyone to view the documents without giving away specific technologies or how they are implemented. This is an important distinction between policies and the rest of the documentation.

Policies are often confused with any written document used to establish governance. These documents do not go into detail about how a piece of technology should be configured or the process of doing so. Again, these are high-level documents that can be developed and given to others. So high level that in fact, you could share these documents without the need for a **non-disclosure agreement (NDA)**.

If the organization were to write a privacy policy focusing on internal staff, this policy would talk about the rights employees have. It could touch upon what systems are being monitored and why. It should not, however, specify the controls used for doing that (which would be the focus of a standard) or the process of how to configure a control (which would be the focus of a procedure).

Topics commonly found in policies include intent for establishing a particular control, a vocabulary, responsibilities of team members, and high-level protections. For instance, an access control policy could include separation of duties, least privilege, the general use of accounts, and a statement regarding periodic reviews. While it states that the company will do this, it does not specify the who, what, where, and why, as well as how it will be done. Providing that level of detail, for public consumption, could put the organization at risk.

For instance, you would not want adversaries to know how you configure an Apache web server, nor would you want others to know how you respond to an incident. This level of detail can be damaging to the organization. If an incident response procedure were to get into the wrong hands, they will know how you intend to respond to a given incident.

Standards

Standards are medium- to low-level documents that depict how an IT resource is configured or what acceptable technologies to use. These documents describe IT resource configurations, the type of encryption to use, or why a user or service account is configured in a particular way. Standards are used to back up what was previously stated in the policy but in greater detail.

A standards document does not go into detail on how to configure a particular setting of the IT resource, only that the setting should be configured. For instance, a standards document would tell you which encryption standards are appropriate or which ones should be disabled. The standard should not detail how to configure encryption on a particular IT resource; that is the job of a procedure. Another example would be a password standard where you state the length and complexity of the password; the following is an example:

- Passwords must be at least 10 characters

- Passwords must meet three of the following four criteria:

  - Uppercase

  - Lowercase

  - Number

  - Special character

- Passwords cannot contain your name or username

As this document states the various types of configurations that are appropriate for an IT resource, it should not be shared outside the organization. Standards detail the services that the organization uses and how they are configured. These details could potentially harm an organization's cybersecurity posture if they fall into the wrong hands.

Providing this level of detail to the public, without an NDA, would be detrimental to the organization. If an encryption standard stated that you still had to use an MD5 hash or SSL v.3 (I hope not!) for legacy systems, then an attacker would go after those systems first. Therefore, having an NDA in place with the person or organization who will be reviewing these documents is crucial.

Further considerations when sharing documents would be digitally signing and adding watermarks. These security configurations ensure that the document was shared and secured by a third party. This also allows tracking a document if the document is accidentally lost or stolen.

Procedures

Would your co-workers know how to perform your job if you were to win the lottery and leave? Would a junior analyst understand how to add a new user to Active Directory if you ever decided to take a vacation? This is the job of a procedure. Procedures provide detailed information on how to perform a particular job function. These must be written in such a way that they provide step-by-step documentation for how to add a new user to your Active Directory, for example. Writing procedures is an art. There may be configuration steps that are second nature, that just you know how to do, but the person reading the procedure may have no idea how you got there. They must be detailed with every step of the process laid out.

Much like how a standard backs up a policy, procedures back up standards. Providing documents focused on how a job function is performed without getting into the details of what is acceptable is the job of the standard. Procedures should be restricted and only allowed to be viewed by employees of the company or a specific department. A diagram for how PSPs flow can be found in *Figure 4.4* later in this chapter.

Policy workflow

There is still more work to do now that we have our PSP drafts written down on paper (or the back of a napkin). As the drafts are documented, you must ensure that they are being reviewed and approved with executive backing. This means that a policy steering committee should be formed to review all policy documents before submitting them to an executive. Policy documents should be broken up into separate sections to ensure that they are easy to read and comprehend.

As previously mentioned, policies are high-level documents that are written in such a way that they can be shared with others without giving away too much information. These policies are given a unique number as well so that it further separates the documents into multiple control objectives. Standards are then provided a number that depicts how they should align with the policy. For instance, if a policy is given 105.00 as a number, then the standard that was written to align with that policy would be given 105.01, and so on. The same is true for procedures as those align with the standard. The procedure in this example would be given 105.01.01 as a number.

Policy documents must be reviewed on a given cycle but reviews should be no more than 2 years apart. This ensures that the documents are kept fresh and up to date. Even if no changes had to be made to the document itself, it still must be reviewed and approved by the committee and signed off by an executive.

With that in mind, let's look at how a policy is approved, created, reviewed, and laid out in more detail.

Getting executive sign-off for policy documents

Executive backing is important for many aspects of your cybersecurity program; policy development is no different. *Figure 4.1* depicts how PSPs flow from the bottom up:

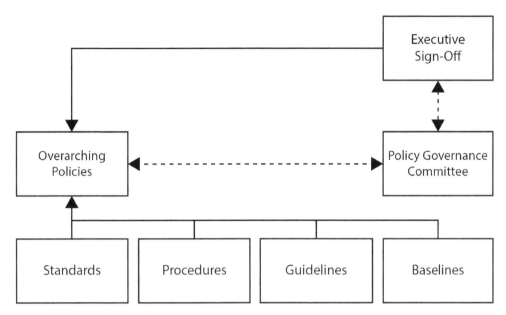

Figure 4.1 – Policy workflow

Standards, procedures, baselines, and guidelines all flow up to an overarching policy. These documents are then reviewed by a steering committee. This committee is required to review all the documents and provide feedback for any modifications deemed necessary.

Everyone in the company should feel empowered to develop policy documents. However, this does not mean that the documents have *teeth* and can be enforced. Once the documents have been presented to the committee, they must then go to the CEO, or a proxy, for official sign-off. This is an important step that should not be overlooked. Any employee can write a document. Without going through a committee and being signed off on by a member of the C-suite, that's all it is – a document without enforcement. When a member of the C-suite signs off, it tells the rest of the organization that the document aligns with the business objectives of the organization.

Creating new policies

All new policies start as drafts. Once the policy document has been written, it is presented to the policy committee. If the draft is denied by the committee, it goes back to the document owner for additional edits. Once it has been approved by the committee, the document is then presented to the CEO or their proxy. *Figure 4.2* depicts policy development and sign-off:

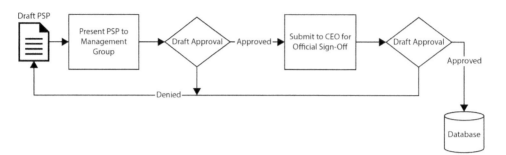

Figure 4.2 – New policy development

If the CEO denies the document, the document then goes back to the owner, who modifies it, then it goes through the committee again before going back to the CEO. The CEO also has the power to deny the document entirely as it may not align with business objectives. When the CEO approves the document, it should go to a document repository. This repository should be open to anyone in the organization to review.

Reviewing policies

Writing PSPs is only part of the workflow. Auditors will need to look at how the policy was constructed and how it is being used. Auditors will also want to know how often the documents have been reviewed. Reviewing policy documents is a natural part of the governance program. They must be reviewed at a given interval (no longer than 2 years) to ensure that they still align with business goals and objectives.

As a policy comes up for renewal, the established document is used as a new draft for the renewal. Any edits must go through the same cycle as a new policy would, first going in front of the steering committee and then being presented to the CEO for sign-off. However, what if the document does not require changes? It must still go through the committee and to the CEO's desk. In IT and information security, things change – almost by the minute. You must get multiple eyes to look at the document to ensure nothing has been missed. The following diagram depicts how the process flow should work when reviewing and renewing a policy document:

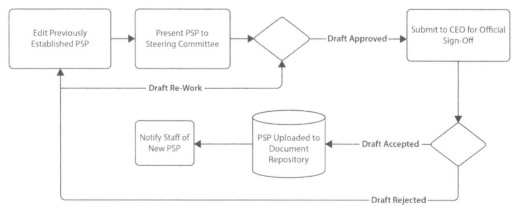

Figure 4.3 – Policy review

As an example, the Heartbleed (CVE-2014-0346) and POODLE (CVE-2014-3566) vulnerabilities were disclosed in 2014 and took advantage of weaknesses found in the OpenSSL library. Part of the mitigations to these vulnerabilities was to disallow or deactivate the SSL/TLS protocol for anything less than TLS v.1.1. When the review cycle for 100.03 – Encryption is due, the standards and procedures should be changed to reflect the requirement. The standard should be changed with a statement saying, "All systems that use SSL/TLS v1.1 or below shall be deactivated," while the procedure for Apache – 100.03.01 should be changed to reflect how to disable weak ciphers in the Apache configuration file.

Building a framework layout

All too often, policy writers combine their PSPs into a single 20- to 30-page document. This makes searching extremely difficult, especially during an audit or a crisis. To ease the burden, look at breaking up the PSPs into smaller documents. This is where building a framework to lay out the documents is important. In the following figure, we can see how the policies flow into standards and how standards flow into procedures:

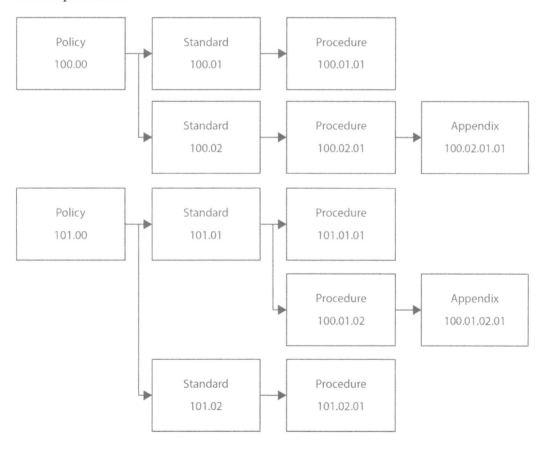

Figure 4.4 – Framework layout

We previously discussed how procedures back up standards and standards back up policies. The layout for the documents will follow the same methodology. The following table depicts how PSPs can be laid out:

Policy No.	Policy Name	Standard No.	Standard Name	Procedure No.	Procedure Name
100.00	Information Security				
		100.01	Acceptable Use		
		100.02	Storage of Sensitive Information		
		100.03	Encryption		
				100.03.01	Apache
				100.03.02	IIS
				100.03.03	MySQL
				100.03.04	SSH
101.00	Asset Management				
		101.01	Hardware Inventory		
		101.02	Software Inventory		

Table 4.1 – Policy framework layout

As the table shows, there is only one policy to the multitude of standards and procedures that have been crafted to back up the policy. The numbering is used to tell us which procedures follow which standard, and which standards follow which policy. As an example, we could have two different procedures for Apache. Each procedure pertains to two different standards. One could be for setting up encryption and the other procedure for Apache could be written on how to configure the Apache service on an Ubuntu server. Both procedures pertain to the Apache web server but align with two different standards.

Exploring policy objectives

To manage risk within the organization, you must set expectations for how the security team will meet or exceed the implementation of safeguards and mitigations. The NIST CSF and CIS CSC list many of the control objectives that should be implemented within the organization's IT infrastructure. These control objectives are what external auditors will assess you against and provide a GAP analysis of findings for the controls that are lacking.

In the following sections, we will look at the control objectives that you should have for your governance program. This list combines many of the control objectives you would find in the NIST CSF and CIS CSC.

Information security

The information security control objectives include being able to protect data. This section includes standards that cover how to dispose of digital media, encryption, storing sensitive information, and the company's **acceptable use policy** (**AUP**). This includes how to successfully destroy digital media so that someone cannot recover files from a disk. It provides answers to the following questions:

- How can we safeguard data using encryption and what algorithms can we use?
- How can we protect sensitive data when it resides on an IT resource?
- What is and is not acceptable to do on a company-owned machine?

Asset management

You cannot protect what you cannot see. Therefore, asset management is considered the most important part of your information security program. The policy should state what asset management is, why it is important, definitions or **configuration items** (**CIs**) that should be stored, and how it positively impacts the business. Standards go into detail as to the various CIs that must be recorded, where they are to be stored, and any thresholds or dollar amounts that an asset is worth to be recorded.

Human resources

Onboarding new employees can be an exciting time for both parties involved. This is also the company's first opportunity to ensure that the new employee understands what is expected of them, especially in terms of cybersecurity. The information security department should work with **human resources** (**HR**) to understand the employee onboarding process. The HR representative should introduce the organization's PSPs to the new employee. Examples of HR policies and standards include the following:

- Employee onboarding
- Employee transfers
- Employee offboarding

Physical security

Closed-circuit television (**CCTV**), fencing, physical barriers in front of the building, mantraps, and **radio frequency identifiers** (**RFID**) badges for door entry are all examples of physical security. Physical security can often be overlooked but is just as important as computer or IT resource security. These must also be documented and signed off on. There are plenty of regulatory requirements (such as Internal Revenue Service Publication 1075) that require physical security to be in place.

Access control

There are quite a few types of access control. You will have to determine which one(s) you want to follow to ensure the security posture of your environment. While role-, discretionary-, and attribute-based access control mostly pertain to users, rule-based access (explained more in *Chapter 9*) applies to access control lists mostly found in firewalls and networking equipment. You should also remember how those are handed out to individuals by following the principles of least privilege and separation of duties.

Incident management

Incident management and response should contain a detailed list of instructions to use during and after an incident has occurred. There are two different types of cybersecurity incidents:

- **Precursor**: A sign that an incident may occur in the future
- **Indicator**: A sign that an incident may have already occurred, or the organization may be experiencing the incident at the moment

What you do and how you react to the incident must be spelled out so that everyone understands what is expected of them. More on risk and incident management will be provided in *Chapter 6*, and *Chapter 7*, respectively.

Business continuity/disaster recovery

What would you do if a disaster hit your company? How would you and your team respond to such an event? Would you know what your most important assets are and how you would bring them back up?

There is a lot of planning that goes into developing a business continuity and disaster recovery plan. It can take months, even a year or two, to get it done correctly. You, or someone on the business side, must plan for a disaster. You need to know what your critical assets are, the order in which they should be restarted, and how they connect to the network. In addition to that, you should also work with the business to understand what an organizational-level agreement would look like for supporting a given system, and what the uptime requirements are for the system.

Security awareness and training

Security awareness is not the same as security training. According to NIST, security *awareness* is done through observation, whereas *training* allows for focused-based topics that assist the employee in performing a job function. Much like how security is a mindset and occurs year-round, so do security awareness and training.

Security awareness can be performed by displaying banners throughout your organization that discuss cybersecurity topics or through seminars or videos. Training, on the other hand, is focused specifically on a job function, such as how to configure certain technology. There is a difference between the two, though it may seem blurred. Security awareness would include videos of what a phishing email could look like, while training is having an employee learn how to configure the mail system to block incoming phishing emails.

Secure network configurations

This section is meant for secured configurations that deal with networking. These include the types of client or site-to-site-based VPNs and firewall policy configuration. Configurations can include the types of protocols being used, whether you allow full or split tunneling, and whether you require the client VPN to always be on or not.

Secure network configurations can also include firewall rule implementation such as deny by default and allow by exception. They could also include types of ports and protocols allowed to flow through the firewall and which firewall zones they are allowed to flow to. The placement of certain applications based on the risk to the data could also be included in this section.

Secure configuration and management

IT resources, whether they are servers, network equipment, firewalls, and so on, should be hardened against a given standard. These include changing default passwords, including **Simple Network Management Protocol** (**SNMP**) strings, disabling or removing built-in accounts such as the guest account, disabling or removing unneeded system services, and implementing encryption. Standards to assist with system hardening include the CIS controls and PCI-DSS. Organizations that develop operating systems, such as Microsoft, Red Hat, and Apple, also publish standards for hardening their software.

Risk management

Organizational risk management, whether it is physical or virtual, must accompany your security program. There are many risk management frameworks available, such as NIST SP 800-37 or ISO's 27005. Topics for risk management could include the steps to identify risk, how to record risks in a risk register, who should sign off on the risk, and threat analysis. More information on risk management will be provided in *Chapter 6*.

Application security

Companies are realigning themselves to become technology companies rather than what they originally set out to do. Nike has taken steps to begin work in the metaverse by hiring fashion designers who specialize in virtual material design[1]. In 2006, Nike designed a tracker for runners called **Nike+**, which is used to help athletes train for competitions and improve their health. Today, just about every company has moved from its traditional product line to software and application development. This is just one of many ways that organizations can interact with their customer base.

Topics for application development include static application security testing, dynamic application security testing, and software composition analysis testing of your software. It should also include topics such as how to secure your **application programming interface** (**API**) or standards used to write your source code. More information on application development and how to test for vulnerabilities will be provided in *Chapter 11*.

These are just a few control objectives you should have for your policy documents. This is a framework that you can use to organize your policy documents. For further assistance, I recommend checking out *The Open Policy Framework*[2].

Summary

In this chapter, we discussed what governance is and how to enforce it through policy. Auditors will, without a doubt, look for your PSPs to understand the technologies in place, how they are configured, and your stance on an implemented control. This means that if you have a standard for antivirus software and it states how often it should be updated, the auditor will look to verify that it is configured that way.

This also helps to train your IT staff on the proper use of technology. AUPs are written for employees so they know what is and is not acceptable use of company-owned equipment. If you have a written policy that states, "Cannot use company-owned equipment for personal use," that tells the employee not to use their company cell phone to make personal calls. You may also have a policy that discourages employees from using offensive language or describing what is acceptable for social media. If you have a **bring your own device** (**BYOD**) policy, that should be employee-facing and state what may happen to that personal device if it was ever lost or stolen.

I also mentioned at the beginning of this chapter that policy documents will move you up in tiers when scoring against the NIST CSF. To get yourself from tier 2 to tier 3, policy documents must follow this flow and be signed off by someone in authority to do so. Documentation is always the last on someone's mind when working with technology, but it is a necessary evil.

In the next chapter, we will check out the hiring process. This will guide you through how to write a job description for analysts, engineers, and architects, which you will need for your organization.

References

1. Golden, Jessica. *Nike is quietly preparing for the metaverse*: `https://www.cnbc.com/2021/11/02/nike-is-quietly-preparing-for-the-metaverse-.html`
2. The Open Policy Framework: `https://theopenpolicyframework.com`

The Security Team

As the new head of security, you may have never had to hire someone for a position. Sure, you can look at LinkedIn or Indeed to help write a **position description** (**PD**), but how do you know whether the job description is right for the candidate you want to hire? We may try and shoot for the moon, asking for years of experience for first-time job hunters or looking for someone with 15 years of Kubernetes experience. Fun fact – Kubernetes has not been around that long as of the time of writing.

Cybersecurity was not taught as a major subject area when I was in school. It was difficult to find a single class that was devoted to security unless you asked the professor. Today, many colleges and universities teach the subject but not to the degree that would benefit a first-time job seeker. There are almost 2 million unfulfilled cybersecurity jobs to date. We have to create positions within the company to which candidates will want to apply and not feel discouraged by. In this chapter, we are going to explore several different ways that you can set your organization up to get the right cybersecurity talent needed. We'll see what the different types of positions are that you should be hiring for and how you can structure the cybersecurity workforce for your company. We will also look at the various skills and frameworks that you can use when structuring your team.

In a nutshell, the following topics will be discussed:

- The need for more security professionals
- The NIST NICE framework
- The different roles in cybersecurity
- Cybersecurity architectural frameworks
- Insourcing versus outsourcing
- Structuring the cybersecurity team

The need for more security professionals

The cybersecurity workforce market is hot right now and it has been this way for quite some time. According to Gartner, in 2017, a senior research analyst with the company stated, "We're as close as possible to our unemployment rate being zero." That unemployment rate pertains to cybersecurity professionals currently working in the field or newcomers trying to break into it. While at the time of writing this book, there was an estimated 348,000 open security positions, the article predicted that by 2022, there would be an estimated 1.8 million unfulfilled positions.

So, how did we do? In 2022, ISC2 conducted a study on open positions in cybersecurity. In the report, they state that the total cybersecurity workforce population is estimated to be around 4.7 million. However, in the same study, ISC2 states that there are still a whopping 3.4 million unfulfilled positions worldwide. How can that be? If it is such a hot market, then why do we have so many unfulfilled positions?

The cybersecurity market can be challenging, yet very rewarding. It can be challenging in that security teams seem to be up against an insurmountable number of cybersecurity threats. Teams become burnt out by the large amounts of pressure placed on employees to always be vigilant. Employees are expected to always be on the lookout for the next threat that could bring down a system, the entire network, or the whole organization. While pay and benefits for cybersecurity positions are also at an all-time high, so is the number of those leaving the workforce. In its third annual Voice of SecOps Report, Deep Instinct reported that 45% of those surveyed had considered quitting the cybersecurity field altogether.

It can also be extremely difficult for newcomers or those just graduating from college to get into the field of cybersecurity. Many entry-level positions still require 5 or more years of experience with knowledge of infrastructure, software development, governance, risk, and compliance, as well as cloud architectures. These are skills many students are not exposed to when graduating from higher education. It is also extremely difficult for those who are new to the field to gain the required experience on their own. While building home labs can cost a few hundred to thousands of dollars, cloud computing can be just as expensive.

Applying NIST NICE framework to your organization

NIST created the **National Initiative for Cybersecurity Education** (**NICE**) Workforce Framework for Cybersecurity (SP 800-181r1). We will use this framework to lay out what an organization, especially its hiring manager, should look for in candidates for particular roles. This framework can also be utilized to create PDs to correctly align candidates with a role within the company. First, however, we have to establish some key phrases that will be used throughout this section:

- **Task statements** describe what work will be performed by a candidate. In other words, a task is defined as something that an employee will perform in order to meet organizational business objectives. For example, a task could be related to configuring network equipment or configuring the Apache service on a Linux server. Task statements should not be confused with knowledge or skills statements as those achieve different objectives.

- **Knowledge statements** differ from task statements in that knowledge is used to perform a task from memory. Knowledge statements could include knowledge of Cisco IOS or how to thwart certain types of threats against an organization. Knowledge statements can also define particular experiences from previous employment, such as how long a candidate has been in a particular field. These can include entry-level versus architectural or even managerial expertise. This can also be a many-to-many relationship in that we can have one or many knowledge statements for a particular task and vice versa.

- **Skill statements** are those that a candidate uses to demonstrate that they can perform a task. For example, a skill would be to configure a pfSense firewall for high availability or to recognize alerts that come from a **security information and event management (SIEM)** system. It could also be used in post-incident handling, such as performing an after-action review or root cause analysis.

Task, knowledge, and skill (TKS) statements can then be put together to create a PD or evaluate a candidate for a cybersecurity position. For instance, the candidate must have 2 years of experience in cybersecurity. The successful candidate must have a background in networking and the ability to configure Cisco IOS. That statement establishes three separate requirements used to align a TSK, as follows:

- **Task**: Configure Cisco IOS

- **Knowledge**: 2 years of experience

- **Skill**: A background in networking

As the hiring manager for the candidate, you must evaluate their competency to fulfill the requirements of TSK statements. Competency is a way to assess the candidates to ensure they know how to perform the job they are applying for. This is much like a final exam for a student, where they must complete the objectives necessary to finalize a task. Evaluating the competencies of a candidate can be performed in many different ways. However, the most common is through **question and answer (Q&A)** or simulation.

Most of us know about the Q&A portion of an evaluation. The candidate sits in front of a committee and is quizzed on concepts they should know. These can be in the form of scenario-based Q&As, specifics on how they could configure a system for security, or through *ice-breaker* questions. Ice-breaker or behavioral questions are more in the form of trying to get to know the candidate and see whether they will mesh with the company's culture. I particularly like these types of questions as you can get to know who the candidate is rather than what they can or cannot do. You also get to draw out their soft skills, understand how they handle situations, and gauge their level of *drive* to better themselves.

Simulations are also a favorite of mine, but they take time to set up. These types of quizzes place the candidate in front of an IT resource and require them to either configure the resource or fix any issues that have come up. Simulations are great when trying to gauge whether the candidate has the know-how or the skill to troubleshoot an issue correctly or see the steps they may take to resolve an issue. These types of questions can be as simple or as complex as required. I have been in troubleshooting scenarios where I had access to the internet to look up commands to resolve an issue. I have also had simulation questions where I only had access to a **manual page (man page)**, which is a piece of documentation about a particular Linux command.

The NICE framework also has a mapping tool that is used to assist employers in developing PDs. The mapping tool can be used to assist with determining the knowledge, skills, and abilities of the candidate. There are a series of questions that the tool will ask when assisting in the creation of the PD. Once complete, the hiring manager can then print the results to be used to create an open position.

You must remember that NICE, much like the CSF and CSC, is a framework meant for guidance. You may have different knowledge or task requirements for the position you are posting for. Hence, you should use the NICE framework to help with writing the responsibilities for the various roles that you want to hire according to your requirements.

Exploring cybersecurity roles

In the cybersecurity field, there are quite a few choices when it comes to different roles. There is more to it than just having a head of security or a **chief information security officer** (CISO). You can have a variety of analysts, engineers, architects, compliance specialists, and penetration testers. Typically, there is not a *one-size-fits-all* type of category. However, to ensure you pick the right talent, you must know what or who you are looking for.

Cybersecurity analysts

Cybersecurity analysts can be your first line of defense when it comes to protecting your organization from adversaries. Analysts perform many of the day-to-day operations that are part of cybersecurity. They are responsible for monitoring the organization's networks, monitoring vulnerabilities, and working with the rest of IT to ensure the security of IT resources. Analysts may also be responsible for performing penetration tests, which are used to evaluate the organization's security. Analysts are also capable of troubleshooting incidents if and when they arise. If the organization experiences issues with a firewall or an IDS/IPS, the analyst is there to resolve the issue.

There are many different types of analysts. A security operations center analyst is responsible for monitoring the network and systems for IT resource vulnerabilities or technological threats against the organization. They monitor spam and phishing emails directed toward your employees to fight off business email compromises that result in account takeovers. SOC analysts may also be responsible for reviewing IT resource vulnerabilities as they are responsible for monitoring security threats.

Those who work in a SOC may also be responsible for threat hunting. Threat hunting is the art of searching for potential indicators, which can point to the compromise of a system. This is a specialized field in which the SOC analyst will need to write queries against a SIEM, log correlation engines, or a system that is collecting events from all IT resources throughout the network.

Cybersecurity network analysts are those who are responsible for ensuring the network is as secure as possible. They are responsible for monitoring firewall rules and ensuring proper logging has been configured. Cybersecurity network analysts are also responsible for reviewing and making suggestions in terms of the secure configuration of devices. Cybersecurity analysts do not, however, make changes to firewall rules or network configurations. That responsibility falls to the networking team.

A DevSecOps analyst fulfills much the same role as a cybersecurity network analyst. DevSecOps analysts are responsible for monitoring the development pipeline and code reviews. These analysts can be there to assist developers in resolving security flaws in various languages. If you have your DevSecOps pipeline set up to block builds for failing SAST, DAST, or SCA testing, they can assist the development team in what they need to do to remediate vulnerabilities in the code.

Cybersecurity threat intelligence (**CTI**) analysts monitor feeds from different sources. These feeds provide valuable insight into potential vulnerabilities, malware infections, and cyberattacks against organizations. CTI analysts collect this data and review it to ensure that the organization is secured against threats. For many small to medium-sized businesses, those responsibilities for CTI fall on SOC analysts.

Cybersecurity engineers

A cybersecurity engineer takes on different responsibilities within an organization. An engineer builds the network, software, and other tools used to improve the security of the infrastructure. Engineers are the ones that create the roads, develop the automation, or write secure code. The engineer is there to build upon current architectural designs created by an enterprise security architect.

An engineer is also responsible for tier-2 level support. This type of support is reserved for assisting the cybersecurity analyst with troubleshooting and resolving issues. A tier-2-level support engineer can also support the organization's customers. The engineer may be responsible for troubleshooting and fixing issues related to IT resources that may be too advanced for an analyst to do on their own.

Cybersecurity engineers act as mentors for the analysts. This involves training an analyst on how to configure an IT resource or how to troubleshoot a potential problem. Engineers are there to assist in training the analyst to make them better and eventually move up from being an analyst to an engineering type of role.

A SOC engineer is responsible for installing and configuring the various monitoring and analysis tools used by the SOC analyst. Tools can include setting up and configuring a SOC's SIEM, getting servers configured, or standing new ones up. They may also be responsible or be a part of the organization's SIRT, responding to incidents when they arise. Engineers could also be responsible for forensic analysis, though that is typically reserved for other specialized fields within cybersecurity.

Networking engineers build the roads for the local area network and the internet while cybersecurity networking engineers install many of the network security tools in place today. Firewalls, IDSs/IPSs, and network packet inspection tools and appliances are all installed per established policies, standards, and procedures. They are also responsible for auditing the firewall rules to ensure compliance. The configuration comes from the networking team, as per the separation of duties laid out in the business and regulatory requirements.

A cybersecurity DevSecOps engineer is there to help build security into the pipeline. Therefore, the engineer is responsible for setting up and maintaining the SAST/DAST/SCA tools used to scan applications for vulnerabilities. They create YAML files, which are typically used in scanning source code, and assist with reporting vulnerabilities. They may also assist younger and more seasoned software developers in resolving errors or fixing security vulnerabilities in the code.

Cybersecurity architects

Architects are the third tier. They design the roads and software to be used by taking business requirements and turning them into goals and objectives. Architects create technical standards and procedures documents, which are used to back up reference architectures. They are also responsible for reducing technical waste in the environment by standardizing the technology used by an organization.

Analysts monitor, engineers build, and architects design the environment. By using business objectives handed down from management, an executive team, or the board, they design solutions to be used by the organization. Through **proofs of concept**, architects implement various types of technologies in the environment to ensure the technology meets the business requirements. Once complete, architects write up implementation documents for the engineers to place the technology into the environment.

Architects are responsible for reducing or eliminating technical waste in the environment through standardization. Standards can come in the form of ensuring the organization simplifies the environment by picking one type of technology for one or multiple reasons. For instance, the organization may standardize on Juniper networking equipment while they use pfSense firewalls and Avaya for **Voice over Internet Protocol** (**VoIP**). Alternatively, architects could decide to standardize on using Cisco equipment for everything.

Standardizing technology helps eliminate using various types of technology in one environment. It also helps architects, engineers, and analysts know and understand two or more types of technologies well. This also makes hiring simpler as it is more specialized. Hiring an architect proficient in multiple technologies may cost you.

In *Chapter 4*, we discussed policies, standards, and procedures and how those are created. Architects may have the responsibility of creating these documents for use by the organization. When an architect standardizes on a particular technology, these requirements must be placed into a standards document. Governance and standardization help other parts of the business understand the various types of technologies that are supported by IT.

Architects are also responsible for tier 3 technical support. In the instance that an engineer has problems troubleshooting technical issues, an architect can be pulled in to evaluate the situation. If the implementation document does not seem to answer all the questions about how to configure and implement the technology, it goes back to the architect so they can figure out the issues or look at a different technology that best suits the needs of the organization.

Cybersecurity compliance specialists

A cybersecurity compliance specialist is responsible for evaluating IT resource security. The compliance specialist reviews the IT resource to ensure that it meets the business and compliance requirements. They are also responsible for creating and/or updating the systems security plan and associated risk register.

The compliance specialist's main responsibility is to perform audits against IT resources within an organization. The audits should review how the resource or technology aligns with a given cybersecurity framework, such as the NIST CSF, and ensure it meets organizational objectives. They maintain systems security plans for all IT resources to ensure they are up to date with the latest information. The specialist is also responsible for working with internal and external stakeholders to ensure the organization is meeting its obligations.

The compliance specialist works with external stakeholders as well. When an external third party wants to evaluate the security and compliance of your organization, the specialist works with them to answer any questions they may have. This also entails requests from second and third parties to fill out security questionnaires that review your organization's security posture.

The specialist is also responsible for maintaining the risk registers for IT resources. As risks are identified, they are raised to management and the executive team for awareness and remediation. They maintain the register and work with IT and the project management office to resolve identified risks.

Head of security

The head of security can take on many different titles. However, it is the responsibility of the head of security, or the CISO, to lead the security program for the organization. The head of security is there to lead, and be led. They should have the mindset of being a *servant leader* who wants to serve people first, leadership coming second.

As we have stated in previous chapters, the head of security is responsible for the overall security program for an organization. It is their responsibility to develop the strategic roadmap and implement that roadmap for the program. The head of security must also ensure that employees perform the work that is required of them and allow their team to grow both personally and professionally.

Leaders want to make the team and the organization better than when they first came on board. This can be done through mentoring, teaching, and offering the ability to better themselves through partnerships and training. Leaders do not need to see the employee in the office to ensure the employee is working, but it is their responsibility to hold themselves and their team accountable when deadlines are not met or the project is over budget.

The head of security is also obligated to provide constructive feedback to their employees regularly. Employees need regular feedback so they know what they are doing right and what they are doing wrong. If a project is not meeting business objectives, you must ensure that the team knows and understands and can get the project back on track.

As you transition from an engineering or architectural role to being head of security, an important factor to remember is that you will have to leave your technical skills behind – at least most of them. This position will require that you know and understand the business, how it functions, what the various product lines are, and how it will impact your department. You will need to learn business speak and how to communicate technical details to groups of individuals that may have never been in IT. You are responsible for creating and leading a technical team and understanding how their roles will help improve the business.

Cybersecurity architects also have architectural frameworks to use at their disposal. Much like cybersecurity frameworks such as the NIST CSF, architectural frameworks assist with creating and managing reference architectures and communicating technical requirements to an executive team. We'll look at cybersecurity architectural frameworks in a bit more detail next.

Exploring cybersecurity architectural frameworks

Cybersecurity architectural frameworks are designed to assist organizations in implementing cybersecurity best practices. Some frameworks will assist you in architecting solutions for a business environment. The **Sherwood Applied Business Security Architecture (SABSA)**, **The Open Group Architecture Framework (TOGAF)**, and **Open Security Architecture (OSA)** are a few examples of such architectural frameworks that can be used. Let's take a close look at them.

SABSA

Many architectural standards can be used. For instance, SABSA [1] is a security architecture framework used by organizations to implement technology securely. SABSA has many different views of security, as depicted in *Table 5.1*:

View	Architecture
Business View	Contextual Architecture
Architectural View	Conceptual Architecture
Design View	Logical Architecture
Construction View	Physical Architecture
Technical View	Component Architecture
Managerial View	Management Architecture

Table 5.1 – SABSA security views

The preceding table displays the different views of a cybersecurity architecture and how you can communicate that architecture. Each view specifies how you would discuss them, depending on the stakeholder involved. In addition to the various views, there is also a SABSA Management Matrix, which reviews the who, what, where, why, when, and how questions regarding how the technology will be implemented and explained throughout the organization.

SABSA allows us to discuss technology and architecture with various stakeholders. For instance, we would not discuss how to configure TLS on Apache with a board member, but we would do so with an analyst, engineer, or architect. Each conversation you have with a stakeholder should consist of business-related requirements when implementing technology, but *how* you communicate those requirements depends on *who* you are speaking with.

TOGAF

TOGAF[2] was built out of the need to develop large system architectures. To standardize how to create these large architectures, TOGAF was developed. First released in 1995, it is now on its tenth version, as of the time of writing in 2022. The framework is based on four domains:

- **Business architecture**: Used to drive overall business requirements

- **Data architecture**: Pertains to the logical and physical data of the organization

- **Application architecture**: Used to describe or provide an overview of the types of applications that need to be developed for an organization

- **Technical architecture**: Used for all the requirements for IT resources, which include hardware and software

TOGAF heavily relies on the concept of the **architecture development method** (**ADM**), which provides guidelines for how to create an enterprise architecture. Like other frameworks, TOGAF and ADM can be applied to the organization or modified to accommodate business requirements. The ADM is based on eight different phases:

- Architectural vision

- Business architecture

- Information systems architecture

- Technological architecture

- Opportunities and solutions

- Migration planning

- Implementation governance

- Architecture change management

OSA

OSA[3] is closely aligned with NIST SP 800-53. It provides patterns of security architectures that can be used that align with this security standard. OSA provides a taxonomy of commonly used words or phrases used to describe what an architecture is and what it could look like. It also provides a control catalog, which is used to apply security configurations to IT resources. The control catalog also comes with information on how it aligns with controls from other security frameworks, such as NIST 800-53, COBIT, and PCI DSS.

TOGAF, SABSA, and OSA are only a few of the dozens of architectural frameworks that can be used for your organization. These frameworks complement each other by adding technical and security-related processes to system design and implementation. TOGAF is so popular that, according to CIO Magazine, by 2016, 80% of Global 50 companies and 60% of Fortune 500 companies used the framework when implementing IT systems and services in their organizations.

Enterprise architects, whether security-related or not, should be familiar with these frameworks or others such as Zachman. Architectural frameworks are an important piece within the overall success of implementing a particular technology or IT resource. This should hold true whether you decide to insource or outsource your workforce, which we will look at next.

Staffing – insourcing versus outsourcing

With talent gaps the way they are in the cybersecurity field, you, as the head of security, have an important decision to make: to insource or outsource. There are plenty of benefits to either approach. It all depends on how you want to grow your team. Growth can be organic, adding employees as the company grows, but what if you are just starting out? Do you look at outsourcing? That can be a possibility too, but it may cost you more initially.

There are plenty of organizations that state they do not have enough cybersecurity professionals and it is hard to hire the right ones. Outsourcing is one way to grow your team quickly by hiring seasoned professionals to perform security tasks. Companies may also outsource their entire cybersecurity team to a third party due to a lack of talent in or around the organization. For example, you may have a few cybersecurity analysts but need assistance in creating a strategy or setting the direction of the program. This is where outsourcing a **virtual chief information security officer (vCISO)** can help.

An organization can also hire contractors as fractional employees only working for the company a few hours a week. Your team may not need a full-time CISO, so hiring a vCISO for 10 to 20 hours per week may be sufficient. Implementing a new system into the organization? You may need cybersecurity architects and engineers just for the duration of the project while your analysts run with the system at the project's end.

Many organizations work with external contracting companies to find and hire talent. Contracting companies are a nice approach in that you can also have a person work for you on a contractual basis. If and when you decide to move forward with hiring someone, you may be able to contract to hire. This allows you to try before you hire to ensure the candidate is capable of performing the job responsibilities and is a good mesh with the team and culture.

Outsourcing has its dangers, however. If you have a third party that has direct access to your company's network, you should validate its security posture. This includes third and fourth parties as well. I am sure we have all heard of stories where a company has a third party monitoring some part of their network. An example is Target, where an external third party had access to the company's network for monitoring various controls. An individual from the third party was phished and adversaries gained access to not only the third party's network but to Target as well.

In another example, approximately 1,500 customers were attacked with ransomware in July 2021[4]. The cause was a flaw in the software used by the managed services provider Kaseya. The software in question was used to remotely manage the IT resources of its customers. This software fell victim to a supply chain attack and was modified to install ransomware on its users' computers.

Alternatively, money can be tight for businesses, and they may not be able to afford to outsource a cybersecurity program. Many organizations intend to grow IT and cybersecurity teams organically. You may be the lone cybersecurity professional in your organization. However, as the company grows, so does your team. It is during these early stages of potential growth that you must look at your program and the strategy you have laid out to better prepare yourself and your team for success.

If insourcing is your overall goal, and you are starting with maybe yourself or another person on your team, it could take time to staff up properly. However, it could lower your costs in the long run and provide greater flexibility for your staff. In many places, insourcing also accounts for remote staffing.

In March 2020, almost everyone was sent to work from home due to the COVID outbreak. Today, many people still work remotely. More and more organizations are willing to open up and allow their staff to work from home. This has greatly widened the talent gap as organizations are open to hiring from pretty much anywhere in the country, or even across the globe.

Organizations can also start by outsourcing their talent with the option to hire after a specific amount of time. This allows them not only to validate whether the candidate has the right skill set for the position but also to see whether they are a good fit culturally.

As you plan out the types of positions you want for your department, remember there are pros and cons to both insourcing and outsourcing talent. Speak with executive management to discuss various options and what the goals are for your department. Some short-term work, usually 6 months or less, requires that you hire someone for a short time for a single project; this would be best suited for a contract worker. For long-term engagements such as operations, you may want to hire someone permanent for the position. It all depends on how you want to *structure* your team.

Structuring the cybersecurity team

The structure of the security team also plays into how you create your cybersecurity program. In *Chapter 1*, we focused on creating mission and vision statements, including what is important to you and your program by creating pillars. One pillar you should focus on is your people. There is something to be said about being a *servant leader* in that, as the head of security, you must take care of your people first.

Previously, we discussed the various roles you should look at hiring for your organization. There are specialty roles such as cybersecurity network or DevSecOps analysts and engineers. There are also architects, who design the systems to be implemented in the organization. Depending on the goals that you have for your team, and the goals of the business, this may shape how you hire and grow the team.

Small to medium-sized organizations may only need a handful of analysts and engineers on the team. The reason for this is that they may not have enough work for a full-time architect to be on staff. While analysts monitor and engineers build, many organizations outsource the design to trusted third parties as they cannot afford to hire additional expertise.

For example, you are implementing a new **identity and access management (IAM)** system. This is a one-time project that needs to be implemented and you do not have the expertise in terms of staff to implement the technology. You may have a handful of applications that need to be integrated, so it does not make sense to hire a full-time IAM architect for a one-time project. Instead, it may make sense to hire out the design and implementation of the technology while ensuring that your engineers and analysts know how to use the new tools once the engagement is over.

If you work for a large organization, then you may want an architect that is multi-disciplined – one that may know the application and network architecture or has expertise in IAM and Microsoft systems. In larger organizations, there may be enough work to hire an architect or two as new components and systems are introduced regularly. Larger organizations with different business units will need assistance from IT and cybersecurity. As a plethora of projects emerge, you will need to hire someone on staff who is capable of getting into the nuts and bolts of a system in order to implement the new technology.

Training employees is a must to keep them up to date and understand how to implement new technologies. The saying goes, if you think you can implement something cheap, just wait until you have to re-implement it a second time. That means, if you do not have the right expertise to implement it the first time, then you are bound to implement it yet again until you get it right. This will cost you in terms of operational and sometimes capital expenditures. You may also want to work with outside vendors or contractors in addition to having an architect on staff. It is all about knowing your employees and understanding what their gaps are.

As a servant leader, it is your responsibility to ensure that your team has the right knowledge and tools to do their jobs effectively. You are there for your team, removing roadblocks while ensuring that they meet deadlines. Removing roadblocks does not mean only removing obstacles such as personnel issues or blockages to moving forward with the project. It also means that you must ensure that they have the proper skills and training to implement the new technology. A servant leader is there to serve their people, not necessarily tell them how to do something.

Summary

Building your security team can be fun and challenging. First, you need to understand the business goals and even your own goals as the head of security to know how you want to staff your team. As you have learned in this chapter, there are plenty of job titles that companies use to align a candidate with a particular job function.

Analysts tend to be the ones that help *keep the lights on*, whereas engineers build the systems and applications to be used by the company. Those systems and applications are first drawn up by architects and the blueprints they design. All three are necessary for your organization – it is a matter of knowing the skill sets of your employees and the amount of work they do to meet their needs.

Apart from learning how to apply the NIST NICE framework to your organization, we also looked at architectural frameworks like SABSA, TOGAF and OSA which help you apply security best practices to your organization. Additionally, we looked at the pros and cons of both insourcing and outsourcing. Many companies choose to insource their workforce. This provides security and peace of mind; however, it is also hard to do. You need approval from human resources and upper management. Outsourcing is the easiest yet most expensive option as you not only have to pay the contracting company fees for the salary but also a *finder's fee* to the contracting company to find someone who will fulfill your needs.

Over the next few chapters, we will go over how to apply the NIST Risk Management Framework to your organization, how to create an incident response team, and how to make your team and the organization cybersecurity-aware through security awareness and training.

References

1. SABSA: `https://sabsa.org/`

2. TOGAF: `https://www.opengroup.org/togaf`

3. OSA: `https://www.opensecurityarchitecture.org/cms/`

4. Tung, Liam. *Kaseya ransomware attack: 1,500 companies affected, company confirms*: `https://www.zdnet.com/article/kaseya-ransomware-attack-1500-companies-affected-company-confirms/`

6
Risk Management

We take risks every day. For instance, did you commute to work this morning? If so, you took a risk by getting in a car, bus, or train to get there. Studies have shown the likelihood of getting into a car accident is much higher than in any other mode of transport. However, you made a calculated decision about how to get to work. What went through your head when you made that decision? What factors went into the evaluation of the overall risk when choosing your method of transportation?

You should have the same mindset when it comes to how you evaluate information technology and security risks. Would you spend $100 to protect $50? No. On the flip side, would we spend $50 to protect $100? Well, possibly. How do we evaluate the risk of losing the $100? If our expected loss of that $100 could occur within days or months, then we would more likely spend the $50 to protect ourselves. If the expected loss could be over 10 years, we would be more likely to accept the risk.

This chapter will review the different types of risks we encounter. Once these risks have been identified, how do we perform quantitative or qualitative evaluations to determine whether we accept, transfer, mitigate, or avoid them? This chapter will also take a look into the NIST **Risk Management Framework** (**RMF**), how to determine qualitative **security categorization** (**SC**) using NIST's **Federal Information Processing Standards** (**FIPS**) 199, and the **systems security plan** (**SSP**).

In a nutshell, we'll be covering the following topics in this chapter:

- Why do we need risk management?
- Exploring IT risks
- The NIST RMF
- Applying risk management to IT resources
- Documenting the SSP

Why do we need risk management?

In life, you take risks every day. When you invest in the stock market or cryptocurrency, you take the risk of losing your hard-earned money. You invest in the hopes of retiring early by putting money into a 401k or into a pension (wait, do they still have those?). You take risks by driving in a car or flying in an aircraft. There are legal risks if you get into trouble and safety risks if you decide to ever parachute out of an airplane or go cliff jumping.

There are also plenty of supply chain risks as well. During the COVID-19 pandemic, computer chip shortages affected millions across the globe as people working from home purchased computers and office equipment to replicate their office settings. Chip shortages have also affected the automotive market, pushing the sale of used cars through the roof while new car sales have been hard to come by.

In early 2020, at the beginning of the pandemic, SolarWinds, a popular network orchestration and data collection tool used by thousands, was hit by a cyberattack[1]. That single attack would then go on to affect close to 18,000 companies, potentially allowing malicious actors to gain remote access to their networks. While authorities believe the original target was the U.S. federal government, according to Business Insider, 80% were nongovernmental organizations, including Fortune 500 companies.

The following year, Colonial Pipeline was attacked by ransomware. In response, Colonial Pipeline had to shut down its systems to prevent the spread. However, it widely affected the eastern part of the U.S., as the company provides 45% of gasoline, diesel, jet fuel, and military supplies (Osborne)[2] to this region. Customers were lined up for blocks trying to get any remnants of gasoline they could to fill their vehicles. While the Colonial Pipeline attack may not seem to be cybersecurity-related, the entry point proves otherwise.

Adversaries attacked Colonial Pipeline's **virtual private network** (**VPN**), allowing attackers to gain access to the company's network. This allowed the perpetrators to launch their ransomware against the internal systems, crippling Colonial Pipeline's ability to perform its job. A compromised username and password were all it took to negatively affect millions of U.S. citizens.

Risk management programs are used to evaluate important topics that affect an organization's functionality, especially around cybersecurity. Risks such as those mentioned in this section should be evaluated to identify and restrict any weaknesses discovered in IT administrative and technical controls.

Exploring IT risks

As you can imagine, there are plenty of risks involving IT and information security. As previously mentioned, supply chain attacks can prevent jobs from being performed. However, although supply chain attacks exist, they are just part of the overall threat landscape. Risks to IT typically come in three categories:

- Human
- Technical
- Environmental

Let's look at these three categories in detail.

Human

Early on in my professional career, I made a mistake. You see, I was told to provide root-level permissions to a few developers in the company, which I had refused to do for months. Having given in, because, you know, you need money, I said that the only way I would do so is if I could teach them what not to do. As you might imagine, it did not end well. While going through my list of commands I said, "now, please never do this…" and as I typed "*rm -rf /*," muscle memory kicked in, and I pressed *Enter*.

Through dumb luck, although this was a critical asset for the company and I had just wiped the hard drive, it was in a cluster of five other servers, so no harm, no foul, and it turned out to be a learning experience for both myself and the developer who learned how to compile and install Linux software.

The human element will be one of your biggest threats. People, myself included, make mistakes. They accidentally delete a file on a file server or remove a record from a database. They are susceptible to spam and phishing attacks. We also have insider threats. Employees who steal software for home use. Salespeople who take their Rolodex with them when they leave. Kids, and some adults, find hacking scripts on the internet and decide to *test* them out. You know, for educational purposes only. You even have the rogue employee who pushes changes to production, during the day, without testing, and says, "huh, well that didn't work!"

The human element is a risk that we cannot live without. If we did not have humans in the workplace, what would we as security professionals do? We need people to use technology safely and securely. How do we ensure this? We have security awareness and training programs, and we teach people via policies, standards, and procedures how to use a piece of technology. We implement security configurations to restrict a user's ability to view sensitive data or remove their capability to delete data. We make multiple backups of the data we need in case something happens.

You will have outside or external human threats as well. These include a single person, a group of people, or state-sponsored threats. While I am splitting them into two, the single person and the group of people will hack for the same cause. They use techniques to disrupt people's lives for financial gain. They send emails stating about winning the lottery or a distant relative leaving a boatload of money behind. I am not talking about those who perform cybersecurity research or ethical hackers who do it for a living.

Then there are state-sponsored hackers who work for governments looking for rival government secrets. The SolarWinds hack is just one example of a state-sponsored hack, as it was believed to originate from Russia. In 2018, 2 Chinese government-sponsored hackers were arrested for breaking into 45 different companies, including NASA's Jet Propulsion Laboratory (O'Kane) [3].

Technology

Does it seem like the refrigerator from 1980 that you're using as your beer (or pop) cooler is holding strong while the new one you bought 6 months ago is already acting up? Machines seem to become more expensive and less reliable. As the saying goes, there are two certainties in life: death and taxes. There should be a third, however, and that would be, *technology fails*. Nothing lasts forever but it is up to you to ensure that the blow is lessened by implementing safeguards.

The **mean time to failure** (**MTTF**) is a calculation used by manufacturers to determine when a component will fail. These components could be hard drives, optical drives, CPUs, motherboards, and so on. Most components will live a long and healthy life whereas others may fail within a few minutes or months. While there is a science to it, it feels random to most.

We must take the MTTF for technology seriously and plan for when technology does fail. Is there redundancy built into the architecture? This includes hard drives in a RAID configuration, multiple servers in a high-availability cluster, or multiple network connections to a switch stack. These features will have to be determined before deploying a new IT resource.

A big concern during the COVID-19 pandemic was our access to reliable internet as very many went home to work. Companies handed out stipends to their employees to help offset the cost of increased bandwidth. This placed even more pressure on **internet service providers** (**ISPs**) to ensure uptime was as close to 100% as possible. Organizations that had a single firewall were purchasing redundant configurations to ensure that their VPNs stayed up and functional.

Planning for technology failures is also a necessity, as we would not have jobs without it. The risks are substantial if not prepared for properly. Planning for system failure may not sound like fun, but it is a necessity. When performing the assessment, one question you should ask is, "how are service level agreements calculated?" Or to put it another way, have you evaluated the risk involved if a system goes down? Do you know how much money your organization would lose if its main website went offline? According to Talent Recap, Amazon loses roughly $443,000 per minute if its systems go down (Agate) [4].

When evaluating a given IT resource, you must take **confidentiality, integrity, and availability (CIA)** into account. How confidential is the data? What would the cost per record be if it were lost or stolen? How important is it to maintain the integrity of the data? For instance, what if someone were to replace peanuts with a different type of food allergy in a medical record? Thus, incorrect information can harm someone if it is not protected and handled properly.

Lastly, availability. In the case of Amazon, losing $443,000 per minute in 2020 may not seem like much when the company made $386 billion in the same year (Kohan)[5]. The percentage of money lost is fractional compared to the total revenue gained. What would happen if your company lost $443,000 due to an outage? Would the company survive?

Environmental

According to `nature.org`, wildlife is second to severe weather when it comes to power failures. Squirrels came in first as the animal that causes the most cybersecurity damage (Ricciuti)[6]. In fact, a website dedicated to providing stats on squirrels causing power outages was set up. It has since stopped collecting stats, but the site is still active (Cyber Squirrel)[7].

You also have to deal with environmental failures internal to your data center. There are data centers across the world. In the U.S., there is no standard place to store your information; however, you have to evaluate the risks. If you were to pick a data center in the western part of the country, you would have to worry about power failures, earthquakes, and potential wildfires. I have seen large, international organizations place their main data center in Miami, Florida (hurricanes???). The desert southwest seems like a likely place but what about severe droughts and power failures – and how does that location play into the people who work for you or the cooling of the building?

To maintain resiliency when expanding your data center footprint, you must look at the environmental conditions of that location. As a rule of thumb, if you have redundant data centers, they should be at least 50 miles apart and not in prevailing winds in case of severe weather. This can protect you from earthquakes, fires, and some weather-related events, although it is not foolproof. Hurricanes have been known to do damage to facilities farther than 50 miles from where they initially hit landfall.

Flooding can also be of concern when evaluating environmental risks. I have seen videos of data centers that have been overrun by water because their IT resources were in the basement. Raised floors will not do you much good if the office windows break due to pressure from standing water from above.

Outside conditions are not the only risks when looking at the environmental element. Conditions must be set internally as well to ensure that IT resources are working properly. Large organizations have teams dedicated to maintaining a data center's environment. They ensure that proper cooling and ventilation occur. They place temperature sensors throughout to better understand heat gaps in their cooling systems. They also ensure there is proper airflow between the servers and networking equipment and figure out how heat can be pumped away from sensitive equipment.

Facebook, Google, and Microsoft have developed new ways to environmentally cool their data centers. Some companies have even begun to develop different types of processors that can withstand higher running temperatures. This has allowed them to run cooling systems at higher, warmer rates than previously. This not only saves on cost by running warmer data centers but it also helps the environment too.

Evaluating human, technological, and environmental risks can take time and resources to fully understand. You cannot take just one type of risk into account when you and your team are making an evaluation. Environmental risks such as weather and nature can wreak havoc on a data center's ability to function. Moreover, the human element of someone accidentally clicking on a malicious link and the failures of IT resources have to be considered too. Therefore, to help evaluate these risks, we will use the NIST RMF.

The NIST RMF

NIST has developed several FIPS and **Special Publications** (**SP**) that discuss overall IT risk to an organization. These include the following:

- FIPS:

    - FIPS 199 – Standards for Security Categorization

    - FIPS 200 – Minimum Security Requirements

- SP:

    - SP 800-18 – Guide for Developing Security Plans for Federal Information Systems

    - SP 800-30 – Guide for Conducting Risk Assessments

    - SP 800-34 – Contingency Planning Guide for Federal Information Systems

    - SP 800-37 – Risk Management Framework for Information Systems and Organizations

    - SP 800-39 – Managing Information Security Risk

    - SP 800-53 – Security and Privacy Controls for Information Systems and Organizations

    - SP 800-60 – Guide for Mapping Types of Information and Information Systems to Security Categories

    - SP 800-128 – Guide for Security-Focused Configuration Management of Information Systems

    - SP 800-137 – Information Security Continuous Monitoring for Federal Information Systems and Organizations

While there is plenty to cover, we will focus on a subsection of documents from the list. First, establish a multi-level RMF for the organization. This will be broken down into three levels: *organizational*, *mission/business process*, and *system*. This is depicted in *Figure 6.1*.

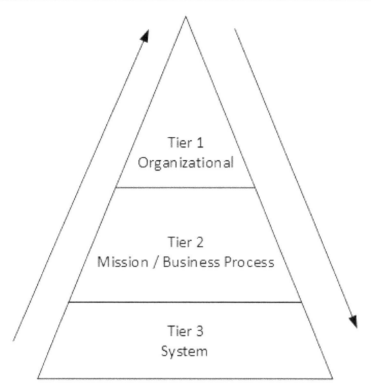

Figure 6.1 – NIST SP 800-37 RMF

Think of the various responsibilities that come into play when evaluating risk. At the top, the executive team is responsible for defining the risk. These definitions then flow down to **Tier 2**, where the definition of risk is put into words through **policies, standards, and procedures** (**PSPs**). These PSPs are also defined in the architecture, establishing standards for how technology should be used in the environment. This can be through written standards, using network and application diagrams, or establishing the types of equipment being used. For instance, does the organization standardize using **OPNsense**[8] or **pfSense**[9] for firewalls? If the **enterprise architecture** (**EA**) has established that only pfSense can be used, then no one in the organization should be allowed to purchase a firewall other than that brand. **Tier 3** then implements the controls that were established in the first two tiers.

Represented by the arrow on the left-hand side pointing up (refer to *Figure 6.1*), if what was established does not work, then there has to be a feedback loop back to the EA or even executive management. For instance, if executive management has stated that the IT organization must reduce their attack surface risk by 5% in the next 6 months, the EA needs to develop a plan to implement a new Layer-7 feature on the firewall. The plan is then given to the engineering team to implement across all regions and all data centers. However, 4 months go by and the engineering team only sees a reduction of 2%, so how can they make up the other 3% in 2 months?

In this case, the EA team must somehow be notified to reevaluate whether mitigations have been implemented as intended. If not, what else can be done? The EA team needs to ask **Tier 1**, or the executive team, "is 5% in 6 months still the goal? Can that number be reduced?" Whatever the executive teams decide will be pushed down through EA back to engineering. This is why that feedback loop is critical to the success of the program.

Next, let's look at the three tiers in *Figure 6.1* in greater detail.

Tier 1 – organizational risk

Organizational risk starts with risk tolerance or appetite. Risk at Tier 1 begins at the executive level and pushes downward as shown by the arrow on the right-hand side of *Figure 6.1*. The executive management team establishes organizational risks by evaluating the human, technical, and environmental risk elements and makes decisions. These decisions are based on the following:

- **Mitigate**: Mitigation of risk is based on the implementation of the controls over an asset. While you can never reach a point of zero risk, mitigation of risk tries to drive the level of risk as close to zero as far as makes business sense.

- **Transfer**: The transfer of risk is when the organization offloads risk to a third party. This can be when using cybersecurity insurance or transferring computer workloads to the cloud. In either instance, you are offloading responsibility for control to another party to accept on your behalf.

- **Accept**: The organization accepts the risk and decides no additional controls are necessary. This could mean that the organization believes it has implemented as many controls as it feels necessary to successfully mitigate a threat. It might also mean that the organization does not have the capacity or resources to mitigate such a threat.

- **Avoid**: There are many reasons why an organization would want to avoid risk. Again, it could come down to cost, or capacity and resources, but it could also mean that the risk is so great that the company just cannot accept or mitigate the risk. Avoiding the risk could mean not implementing a new service or making an IT resource highly available. It could mean that sensitive data cannot reside in or be transferred through a particular resource because the risk is too great.

Tier 1 also introduces the concept of *framing* or scoping the environment. When performing a risk assessment, you must learn how to scope the environment to understand the risk that is presented. In *Figure 6.2*, you can see the server environment for an organization:

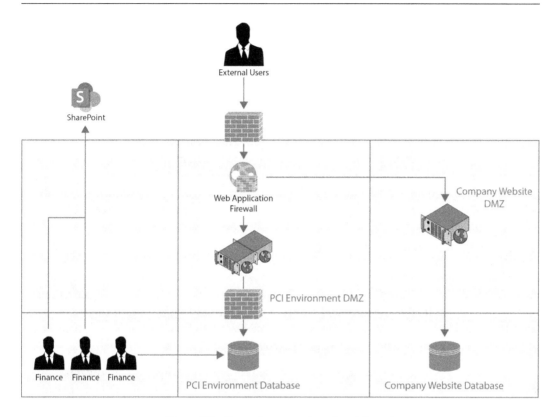

Figure 6.2 – Example corporate network layout

On the left, you have the finance department; in the middle is the company's PCI environment; and on the right, you have the public-facing website. External users have access to the website of the PCI environment to purchase goods from the company. The public website only contains public, non-sensitive information. The PCI database stores credit card information, transaction history, and **personally identifiable information** (**PII**) about the customer. All inbound traffic from the internet must pass through an external firewall, a web application firewall, and then the web server. Database transactions must also travel through a firewall before the web server can access the data.

This diagram depicts how to segment the environment, keeping the PCI environment separate from the public web server traffic. While portions of the environment are separated by the firewall, the DMZ firewall itself spans two different scopes. First, it filters traffic coming in from the internet, only allowing ports 80:tcp/443:tcp to access the web servers, the PCI, and public-facing websites. Then, it passes traffic to the **web application firewall** (**WAF**) and then onto the web server. The DMZ firewall and the WAF both belong to the PCI and public-facing web server environments. When scoping, if there is a shared resource – in this instance, the DMZ firewall and WAF – it not only belongs to non-sensitive information but also passes PCI data as well. Because of this, they are both in scope for PCI.

Say that you did not have the firewall splitting the traffic between the PCI and public web server scopes. If you did not split up the environments with a firewall, both the PCI and the public website would be in scope for the same regulatory requirements. To properly scope the environment, you must first understand how the network is segmented. Knowing which IT resources can communicate with each other helps better understand the risk.

The second part to understand when developing scope is what the interdependencies are, also known as common controls. If you are in a full Microsoft Windows environment, chances are that you have an **Active Directory Directory Services (AD DS)** environment for authentication. It is also a good assumption that you are using Group Policy Objects to manage the security of the servers, laptops, and desktops. Since you are implementing security controls and using AD DS for authentication, this would be considered a common control. It would also be in scope for the regulatory requirements.

Tier 2 – mission/business process

The risk appetite, how much risk you are willing to accept, established at the organizational level directly affects all levels below it. Tier 2 risk management takes what was set at the organizational level and then applies it. The application of a control mechanism to a given risk is carried out via administrative and technical controls. This means that establishing policies, standards, and procedures is as important as the implementation of a new IT resource or component.

Tier 2 defines and prioritizes defining and establishing processes that align with an organization's business goals and objectives. This can be performed through PSPs – however, you must take risk into account. The organization should perform risk assessments on all existing and newly implemented technology to ensure that it meets the risk appetite.

In addition to administrative and technical controls, it also ensures that the organization has established an EA team. This team is responsible for making recommendations for new IT resources within the organization. The team also takes a holistic view of the environment and develops standardizations for the infrastructure. It reduces technical waste through these standards to ensure that current technology is being used as intended. Newer technology is evaluated to see how it aligns with business requirements and whether it duplicates currently implemented technology. The NIST RMF also ensures that security is built into the EA team.

Tier 3 – information systems

Risk at Tier 3 is evaluated at the IT resource level. Engineers are responsible for implementing the risk-based controls for IT resources that were established at Tiers 1 and 2. At Tier 3, controls are selected, implemented, assessed, authorized, and monitored. Although it may seem that these tasks are performed at the engineering level, authorization should never be signed off by someone in EA or an engineer. Authorization happens at the executive level. We will cover more on Tier 3 in the next section.

Applying risk management to IT resources

How we apply risk management and what we apply it to is just as important as creating the RMF. This next section will discuss how to evaluate risks and implement risk management for your IT resources.

Categorize

When performing risk management, you must categorize the level of risk for an IT resource. This will determine how much a resource is impacted by a particular risk. To do so, we must apply a risk indicator to the CIA of the asset and its data.

This phase is based upon the Special Publication 800-37 RMF for Information Systems and Organizations developed by NIST. As we move through the categorization phase of the RMF, we take a qualitative approach to identifying risks based on FIPS 199 and its low-, moderate-, and high-risk categories.

FIPS 199 categories

We will use qualitative analysis to calculate the overall risk score of an IT resource. The following are the thresholds described in FIPS 199. Evaluate each part of the CIA triad and apply a threshold to it by using the following criteria.

Low

A label of **Low** should be reserved for parts of the CIA triad that pertain to minimal loss. Examples could include the following:

- The loss of a single server in a high-availability cluster, causing zero downtime
- Someone receiving a spam email
- Ping sweeps across your external network

Moderate

The **Moderate** threshold is reserved for attacks in which an employee has been manipulated into acting in a way they would not normally. This can also include system downtime, which can affect revenue streams for the organization. Examples include the following:

- Computer virus outbreaks
- Illegitimate wire transfers or the transferal of funds
- A Tier 1 system being down for extended periods

High

The **High** threshold is used when a catastrophic event has occurred in an organization. This could include the following:

- Ransomware attacks
- Cyberattacks against medical facilities
- Substantial loss of monetary funds

More information on the standards for SC published by FIPS can be found at `https://nvlpubs.nist.gov/nistpubs/fips/nist.fips.199.pdf`.

Calculating risk through SC

There are a few different ways you can calculate risk in an environment. One calculation is *risk = likelihood * impact*. This calculation has been used throughout textbooks to depict an overall quantitative score. Once we have determined an overall score, we can rank the risks from high to low. This risk ranking is then used to understand what risks should be mitigated first.

For example, what is the likelihood that an organization will experience an **account takeover** (**ATO**), and what is the impact on the organization? If an ATO is directed at an employee in food service, the overall impact may be minimal. If the ATO affects someone in finance, however, the impact may be much higher depending on what the employee has access to. Another way of calculating overall risk is through the use of FIPS 199 and the SC described in the previous section.

The SC of an IT resource and the data that resides on this resource is calculated based on the low, moderate, and high thresholds previously discussed. In addition to low, moderate, and high, data can also have a threshold of **not applicable** (**N/A**). N/A is only applied to data if there is a potential for the data to be public knowledge or non-sensitive information, such as that found on a public-facing website. N/A is never associated with an IT resource. To calculate the threshold, we use a particular formula.

How secret is the data used in an IT resource? Can everyone view it, or just a few people? Confidentiality refers to how secret the data being used in a system is. When evaluating low, moderate, or high for confidentiality, you must take into account whether a confidentiality breach would harm your organization or possibly your customers.

Integrity is labeled based on whether the data can go unaltered without causing harm to the business or your customers. For instance, you work in a hospital setting and you accidentally enter in peanuts instead of penicillin as an allergen, or that information was altered in transit.

Availability is the uptime for a given IT resource. Can the system go down at all for maintenance or must it stay up 24/7, 365? If it were to go down for any reason, what would the impact be for an organization?

SC data/IT resource = {(confidentiality, impact), (integrity, impact), (availability, impact)}

As an example, we can take the public-facing website we previously discussed during our discussion of scoping and apply impact to it. As the data is non-sensitive with a relatively low impact in terms of integrity and availability, we would use the following formula:

SC public data = {(confidentiality, N/A), (integrity, low), (availability, low)}

SC web server = {(confidentiality, low), (integrity, low), (availability, low)}

Public data = low

Web server = low

Low

We now take the highest threshold or watermark and apply that to the overall score of the resource. In this instance, it would be considered a *low* threshold. What would the score look like if we applied the same method to the PCI environment?

SC PCI data = {(confidentiality, moderate), (integrity, moderate), (availability, moderate)}

SC web server = {(confidentiality, moderate), (integrity, moderate), (availability, moderate)}

PCI data = moderate

Web server = moderate

Moderate

Lastly, if we had a public-facing website with non-sensitive data but we had to ensure that the server was up all the time, our calculation could look as follows:

SC data = {(confidentiality, N/A), (integrity, moderate), (availability, moderate)}

Data = Moderate

Select

Once we have determined the SC of the data and the IT resource, we now must select the security controls. Selecting the security controls to implement for an IT resource is dependent on the categorization phase. When we are selecting the controls, we look at the categorization of low, moderate, and high, and the respective controls. This is normally handled by NIST SP 800-53 in which security controls are based upon the categorization of FIPS 199. However, in our instance, we are applying the NIST Cybersecurity Framework, which does not have this type of classification.

While this applies to a different framework than what we are using, it does not derail the overall objective of choosing our controls. The categorization phase is still warranted, as we must define the overall risk of the IT resource. Risk will never go away – we evaluate the risks to our assets and how we mitigate those risks. The risk that is left over or the residual risk is what we live with. Remember, driving zero risk is both costly and unachievable. We must balance risk with what makes fiscal sense.

Implement

Once you have categorized the risk to your assets and information and selected controls on this basis, you need to implement them. Controls are also evaluated based on acceptance, mitigation, transfer, or avoidance. This includes not only controls specific to an IT resource but also common controls. Again, these are systems that are in place to implement security control for two or more IT resources. Common controls include authentication, antivirus mitigation, encryption, and standardized configurations. These are just a few examples, but you get the idea.

The implementation of controls can be performed through manual or automated processes. Those controls can be pushed down by AD Group Policy Objects or the centralized management of antivirus software. There are open source tools that can be leveraged to auto-provision software and operating systems too. Tools such as Puppet and Chef can be used to automatically change configuration files to get services working as intended. Terraform can also be used to programmatically construct networking in the cloud, automate the creation of servers and storage, or configure DNS settings. Many automation tools state that they can get an environment up and running within minutes, with servers provisioned and the network created.

Assess

After the controls have been implemented, it's time to assess their effectiveness. Do the implemented controls work as intended? Have they been implemented properly? Do they meet the security requirements for the project? We also must ensure that previously configured common controls still meet or exceed the security requirements for the system.

We first look at the SC and the regulatory requirements. Once we understand those two, we can assess whether the risk is within acceptable limits. In addition to assessing the controls implemented on the resource, we must look at common controls too. If a system was originally set up to protect sensitive PII, that data may not be at the same level as a PCI environment. The common controls, in addition to the implemented controls, also need to be assessed. For instance, the firewalls that were shown in *Figure 6.1* may have been set up only to protect a non-sensitive web server and the employees of the company. Now, with PCI, we must implement multi-factor authentication, extend our logging retention standards to 12 months, and ensure that the firewalls are configured to only allow the required traffic into the PCI environment.

If these configuration settings are not set properly, you will fail the assessment. To assess the environment, scope it appropriately. Only include systems that directly affect the security and compliance of the environment. Again, if it is not properly scoped, the entire environment is considered, which can extend the length of time it takes to make an assessment – and cost more.

Authorize

Once we have worked through the previous steps, the next step is to authorize the system to go into production. A system's security plan should be developed and should depict the IT resource and common controls that were implemented during the previous phase. The authorization of an IT resource should come from someone on the business side. Remember, as an information security professional, no matter what stage you are at in your career or how high you are on the food chain, you should never accept the risk of an asset on behalf of the business. This is the responsibility of somebody on the business or executive side.

Security professionals should not be in the business of accepting risks on behalf of the business. As mentioned previously, security is an advisory-type role. Security should be an overlay to the controls and objectives for IT and the business. Someone from the executive team, or the business, should sign off on the appropriate objectives and risks for the business. This is not part of the security professional's job. The responsibility must be reserved for a chief risk officer or another risk executive. In the absence of a risk executive, the responsibility then falls on someone else within the executive team.

This is also why we have data custodians and data owners. Data owners tell data custodians which security controls should be implemented for the data to stay compliant with regulations. They own that part of the risk since they do not fully abide by the regulation. Should a security analyst be responsible for accepting a system that does not comply with PCI DSS? Again, it is your responsibility to explain the risk, not accept it.

Monitor

To ensure the security of your IT resources, you must monitor the security controls. Monitoring of security controls can be done manually or automatically. While we strive to automate as much as possible, you have to rely on manual reviews to ensure the automated scans are performing as needed.

Implementing monitoring controls can take time to implement and sometimes, you may feel that you are not monitoring enough. You need to compile metrics to show the progress of the security program to the board or outside auditors. When performing scans, the first step is to set up **Simple Network Management Protocol** (**SNMP**) on the IT resource. Most, if not all, manufacturers of software and hardware implement SNMP for monitoring. SNMP comes in 3 different versions – 1, 2c, and 3. SNMP versions 1 and 2c are the least secure and default their community strings to public and private. The public strings are used as read-only while using the private string allows someone to write a configuration remotely. SNMP version 3 adds authentication and encryption.

SNMP is mostly used as a way to monitor security baselines. Open source tools such as Nagios[10] are used to monitor and maintain historical data about how that resource meets the baselines. It can also be used to monitor and track security events such as malware detection. This may not come default with the scanning tool and specific **Management Information Base** (**MIB**) strings may need to be scanned to gather that data. Check with the software manufacturer to see whether this, along with many other security-related activities, can be monitored.

Most paid tools also come with automation built in. Antivirus and vulnerability management tools can be configured to scan automatically. Most vulnerability management scanners will perform external scans, looking for open ports to detect their version number against a database. Antivirus applications perform many of the same functions – however, they scan the asset internally to detect malware. Security and hardening configurations can also be scanned using a CIS-CAT scanner. While using automated scans may seem to be the only way to go when monitoring the security configurations and posture of a system, manual observance should also be part of the monitoring phase.

Stuxnet, malware that targeted Siemens' **programmable logic controllers** (**PLCs**), was used against Iran and its nuclear program. A fascinating part of this story was that it took over the monitoring tools used by the employees. As Stuxnet spun the centrifuges up and down, it was also programmed to show that the reactors were functioning normally. The employees had no idea that the nuclear reactor was malfunctioning, as their automated monitoring tools showed that it was running optimally.

The moral of the story is that we cannot fully rely on automated tools. Eventually, configuration drift will occur, or an automated process is bound to fail. You or someone on your team should evaluate the automated controls to ensure they are in proper working order and provide the information that you require. This does not have to be done every day, but it should be reviewed at least quarterly – if not more frequently.

As the head of security, you will need to evaluate the overall risk to an IT resource and how it relates to the overall business objectives. Once risks have been identified, these risks will need to be documented in an SSP. This plan is used as a living document that will need to be continuously reviewed and updated as new risks are identified or mitigated. Let's look at this in the next section in more detail.

Documenting in the SSP

As you work to develop your RMF, you also need to document how an IT resource was developed and configured, what common controls were implemented, and its production readiness. This all comes in the form of an SSP. An SSP is a living document that details everything about an IT resource and how the environment was scoped. This is the document that will be presented to the executive team for official sign-off before going to production.

An SSP is a detailed document that should state who the project champion is, the technical leads or those working on the project, the project manager, and who signed off on the project. It should include their phone numbers, email addresses, and any other contact information about those involved in the project. This is to track who worked on the project in case there are questions after the project has been completed.

In addition to the contact information of those involved, it should also detail the controls that were implemented. This includes security controls of the IT resource, their configurations, common controls, and whether certain risks were identified. Not only should these be documented in written form but there should also be logical network and application diagrams that show the layout of the environment. Dependencies also need to be documented, especially if the SSP covers common controls such as antivirus or AD.

Creating an SSP is much like performing a first-party security assessment. This requires the individual or party to be comfortable with performing risk-based assessments against IT resources. Given that these assessments only capture one point in time, they should be performed periodically. Again, security is a mindset and must be ongoing. It can be reviewed when a new vulnerability such as Log4j is presented. How did you resolve it? What did you use to identify it in the environment? Auditors and the board of directors will want to know the answers to these questions.

When writing an SSP for a new IT resource environment, risks will be identified that also need to be documented. These risks should be resolved before the system goes into production, or the identified risks will lie dormant. All risks should be placed into a risk register for historical tracking purposes.

What is a risk register?

A risk register is also a living document that is used to track the identified risks to an IT resource's environment. The document can be part of the SSP or a completely separate item altogether. The risk register is also a living document of the IT resource's environment. This means that the document is continuously updated based on various identified risks. Once a risk has been mitigated, that risk is never deleted from the register and only closed out. This is to maintain historical information about the resource.

The risk register should contain the following sections:

- **Risk ID**: The unique ID for the identified risk
- **Description**: A description of the risk
- **Priority**: How high-priority resolving this risk is
- **Identifying person or system**: Who or what identified this risk
- **Risk response type**:

  - Accept
  - Mitigate
  - Avoid
  - Transfer

- **Risk response**: What will be done to mitigate the risk or who it will be transferred to
- **Risk owner**: Who is responsible for resolving this identified risk
- **Cost**: Operational costs
- **Duration**: How long it will take to resolve

Risk registers include the identified risk and the chosen mitigation or resolution. For instance, your organization is concerned about the number of ransomware attacks that have occurred throughout the country. One way to mitigate this risk is by ensuring that the organization has adequate backups and restoration procedures.

The organization may also be in fear of ATOs and want to ensure that there is a high level of trust when someone uses an identity for authentication. You could write up a password standard that includes ridiculously long password requirements or enforce the use of multi-factor authentication. You could also require anti-tampering mechanisms to be used to prevent session hijacking. These should be written in the risk register and tracked to ensure they are mitigated according to business objectives.

Driving to a resolution

The last item on the list is, "how are we going to actually resolve or mitigate a risk?" A **Plan of Action and Milestones** (**POA&M**), or simply put, a project plan, is then created. The POA&M is used to show all the bits and pieces of how the organization mitigates a risk. While the risk register designates the risk owner, this is not necessarily the person who will be doing the work.

The project plan will show the steps used to resolve a finding in the risk register. It will show the steps used to resolve the finding, and how long it took with start and end dates. It will denote the responsible individual that will be performing the task, and then its status. As the procedure for the finding moves from beginning to end, the entry will never be deleted, only eventually marked as resolved.

A finding that shows up one year will stay a finding until you can show proof that it has been resolved. An auditor will review previous findings and if it was previously in a report, you will have to confirm that your organization has resolved it. If you delete a risk from the POA&M or the risk register, how will you ever show proof and verify that it has been resolved?

The SSP is, therefore, a living document that is tied to an IT resource from before it was put into the environment and kept after it is long gone. It is used to evaluate the IT resource against a given framework to ensure that it meets the security requirements of an organization.

Summary

We take risks every day. Whether they are human, environmental, or technical, they are risks that we evaluate constantly. When evaluating the risks to an IT resource, the same challenges apply. You must have a structured way to evaluate and respond to risks once they have been identified. In this chapter, you have learned to evaluate these risks using the NIST RMF and apply risk management in various stages from categorizing to monitoring.

As mentioned, once risks are identified, they should be published in a risk register and remediated using the steps defined in a POA&M. An SSP captures all the system components, establishes the scope, identifies the parties involved, and is the document that should be signed off by an executive before a system goes into production. Remember, identified risks and remediation steps should never be deleted from these documents. Once a risk is closed out, it should be marked **closed** in the status column and never be deleted.

Over the next few chapters, we will be looking at incident response, security awareness, and training for your organization, following which we will review network security.

References

1. Jibilian, Isabella. *The U.S. is readying sanctions against Russia over the SolarWinds cyber attack. Here's a simple explanation of how the massive hack happened and why it's a big deal*: https://www.businessinsider.com/solarwinds-hack-explained-government-agencies-cyber-security-2020-12

2. Osborne, Charlie. *Colonial Pipeline ransomware attack: Everything you need to know*: https://www.zdnet.com/article/colonial-pipeline-ransomware-attack-everything-you-need-to-know/

3. O'Kane, Sean. *Chinese hackers charged with stealing data from NASA, IBM, and others*: https://www.theverge.com/2018/12/20/18150275/chinese-hackers-stealing-data-nasa-ibm-charged

4. Agate, Samantha. *Amazon Went Down For 15 Minutes & Jeff Bezos Lost More Money Than An Average American Will Earn In A Lifetime*: https://talentrecap.com/amazon-went-down-for-15-minutes-jeff-bezos-lost-more-money-than-an-average-american-will-learn-in-a-lifetime/

5. Kohan, Shelley. *Amazon's Net Profit Soars 84% With Sales Hitting $386 Billion*: https://www.forbes.com/sites/shelleykohan/2021/02/02/amazons-net-profit-soars-84-with-sales-hitting-386-billion/?sh=3740f80a1334

6. Ricciuti, Edward. *Fear the Squirrel: How Wildlife Causes Major Power Outages*: https://blog.nature.org/science/2019/10/29/fear-the-squirrel-how-wildlife-causes-major-power-outages/

7. Cyber Squirrel: https://cybersquirrel1.com/

8. OPNSense Firewall: https://opnsense.org/

9. pfSense Firewall: https://www.pfsense.org/

10. Nagios monitoring software: https://www.nagios.org/

7
Incident Response

It is 3 am, and you are awakened by your cell phone. It is work. You think to yourself, "This cannot be good!" You answer, and the voice on the other end says the company has been hit by a ransomware attack. Quickly, you spring into action, but the only question is, are you prepared for such an incident? Do your employees know what to do if this or any type of incident hits your company?

You have done everything right. Developed the cybersecurity strategy, started putting projects together to reduce your overall risk posture, and yet you are still hit by an incident. It is OK, it has happened to all of us, and all signs point to the fact that this will not be the only incident you will face in your career. However, you must not only prepare yourself for such an incident, but you must also prepare your team and the rest of the organization.

Much like many things in cybersecurity, incident response is not just a technical problem. Incident response is both an **information technology** (**IT**) problem and a business problem. It becomes a technical problem because it affects IT resources. It is a business problem because the business will want those systems back up and functional as soon as possible. The only way that you can get better is through training your employees. You must *train as you fight*! Meaning you should train staff members using scenarios as close to real-life situations as possible.

There are plenty of incident response frameworks that you can choose for your organization. As this book focuses on the various **National Institute of Standards and Technology** (**NIST**) frameworks available, we will continue down that path in reviewing these. In this chapter, we will discuss the following pain points:

- NIST incident response methodology
- Incident response playbooks
- Train as we fight

NIST incident response methodology

The incident response framework, based on NIST SP 800-61r2, has four main stages: preparation, detection and analysis, containment, eradication, and recovery, and post-incident activity. There are feedback loops between the various stages to provide updates and feedback on how the organization responds to an incident.

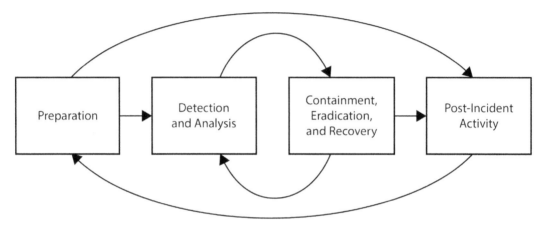

Figure 7.1 – The four stages of the incident response framework

Let us look at each of these stages in more detail.

The following steps are critical as they will prepare your team to identify when an incident occurs. You will be able to determine what the incident was from an analysis of the threat actor. Then you want to contain the incident to as few impacted objects as possible and then eliminate the threat.

This all leads to the post-incident stage and a review of how to improve the process. You need to answer questions such asthe following:

- Was the information provided as quickly as possible?

- What else would have helped in this situation?

Preparation

Preparation for a cybersecurity incident takes practice. To prepare for an incident, you must have a plan in place, decide who will be the incident commander, and pick who will be part of the **security incident response team** (**SIRT**). Your SIRT must include stakeholders from the business, as this is not just an IT problem. Much like business continuity and disaster recovery, an incident response must include individuals from outside of IT. You must have an understanding of which IT resources are critical for the business to function. You must also decide who will communicate with internal and external stakeholders, as this requires someone from marketing or the communications department. You may need to include your legal department if a large or material incident requires reporting to your cyber insurance company or you have to disclose information regarding a breach of information.

Let's look at the factors that come into play during the preparation stage.

An incident response plan

Your incident response plan will come in the form of a policy and multiple standards and procedures documents. As discussed in *Chapter 4*, **policies, standards, and procedures** (**PSPs**) are important artifacts for the organization to create. PSPs establish governance in the organization by explaining to employees what is and is not acceptable when performing work for the business.

To begin preparation, create a policy document that provides a high-level overview of the incident response plan. In the policy document, we shall depict the four stages of incident response and explain their importance. The policy document should also discuss the establishment of the incident response team. This should include a separate hierarchy or organization chart of those who are a part of the SIRT and their contact information. The organization chart must state what the roles and responsibilities are for each member of the SIRT. The following is a list of suggested groups of individuals to have on the SIRT:

- **Incident commander**: The incident commander is a seasoned professional who has been selected to lead in the event of an incident. The incident commander is responsible for the development of the SIRT and policy documents and leads the team during an incident.

- **Deputy incident commander**: The deputy incident commander is second in command. When the incident commander is not available, the deputy will take over their responsibilities. When the incident commander is running the response, the deputy can run the response from another location if the workforce is distributed or hybrid.

- **Networking team**: The networking team is responsible for performing diagnostics on network equipment and searching for possible compromises. The network team is responsible for running network packet captures, opening or closing firewall ports, and implementing additional security measures to block attacks coming into the network from the internet.

- **Help desk**: The help desk is truly the first line of defense. The help desk is responsible for taking incoming calls and disseminating the information to the rest of the team. If a user calls and says their computer is infected or their password was used from an unknown, untrusted location, these could be indicators of compromise.

- **Server administrators**: Server administrators are responsible for troubleshooting server issues. They can access servers, run diagnostic tools, locate malicious files, and review system logs.

- **Security department**: The security department is responsible for many of the network and server analysis tools. The company may have **data loss prevention (DLP)** tools, **intrusion detection** or **intrusion prevention (IDS/IPS)** tools, and possibly a **host intrusion prevention system (HIPS)** or **network intrusion prevention system (NIPS)**. The security department may also have forensics employees to run analyses on IT resources.

- **Legal team**: The legal team must also be included if there is a request for payment of a ransom or loss of personally identifiable information, intellectual property, or other sensitive documents.

- **Communications and marketing department**: The marketing and/or communications department will be your public-facing department. They will be responsible for communicating information to the public and the media.

The plan should reside in multiple places in the event of a catastrophic failure or if the systems on which the plan resides fail. Failure could be the result of a system or network failure, accidental deletion, or from malware, such as ransomware. The incident response plan should reside in a document repository that is either on-premises or in the cloud. If there is a possibility that local or cloud storage could become compromised, you should have the plan in a secondary location. It is also highly important to keep local hard copies for review if every IT resource is down.

Protection of the incident response plan is of the utmost importance too. Not only do you need to protect the incident response plan from accidental deletion, malware infection, or system failure, but the plan must also be protected from prying eyes. Your incident response plan provides valuable insight into how the SIRT will respond to an incident. If an attacker were to get a hold of the plan itself, they would know how the organization would respond to an attack. With this knowledge, the attacker could craft an attack you did not plan for and possibly gain access to the network. If these documents are released to those outside the company, a non-disclosure agreement should be signed by all parties involved.

Communications

As the incident commander, you are running the SIRT and are responsible for ensuring that the teams communicate. Network and server engineers must learn to communicate with each other during an incident. They should call out what they are looking at and what they are doing. This is to ensure that no two teams are looking into the same thing at the same time. When an incident occurs, time is critical. It is critical to stop the attack and eradicate it from the environment. When multiple people or groups work on the same problem, we duplicate effort.

There should be a set of standards for how the SIRT communicates internally and externally. The standards should detail the who, what, where, and when to communicate. They should also establish thresholds for the various types of incidents that you can communicate. For instance, the standards could state that employees receiving phishing emails can be communicated to the public or only to those in the same field. It should also clearly state whether you want only catastrophic incidents to be communicated to the executive team or the board of directors.

Traffic Light Protocol

The **Traffic Light Protocol** (**TLP**) was developed to disseminate sensitive information to others. The TLP was designed to assist the incident response team in sharing information with a variety of people. It is used as a grading scale for how a team or an organization can distribute potentially sensitive information to others. Let's look at what each of the colors signifies:

- **TLP:Red**: Information can only be disclosed during a meeting or by person-to-person communication. This information can be disclosed to executives, the board, or law enforcement. Information with a red designation should not be disclosed outside the organization and only on a need-to-know basis.

- **TLP:Amber**: Information can be disclosed only within an organization and with trusted third parties on a need-to-know basis. Other designations of information disclosure are based on risk to the organization.

- **TLP:Green**: Information at the green designation cannot be disclosed to the public. It is advised to keep information regarding an incident within the organization or the community. The community is defined as those with similar interests or fields of business. There are many **Information Sharing and Analysis Centers** (**ISACs**) that your organization can be part of. These organizations provide valuable information, including cybersecurity-related information, to their members. Those centers include the following:

  - **MS-ISAC**: Multi-state (used by state, local, tribal, and territorial organizations)
  - **IT-ISAC**: Information technology
  - **Auto-ISAC**: Automotive
  - **FS-ISAC**: Financial services
  - **REN-ISAC**: Research and education network

- **TLP:White**: Information designated at TLP white can be disclosed without restrictions.

Call tree

Another necessity is having an up-to-date contact list or **call tree**. Much like on-call rotations, you must maintain a call tree to have the correct contact information for your SIRT. The call tree should contain at least names, phone numbers, email addresses, and titles. The SIRT organizational chart depicting how the various teams line up for reporting should also be included with the call tree.

The call tree also includes contact information for outside counsel, local law enforcement, cyber insurance, and other pertinent information. While this information may not be needed during an incident, it will be needed once the response has concluded. If you are part of the **Federal Bureau of Investigation's** (**FBI**) InfraGard[1] or your organization is a member of the FBI's **Domestic Security Alliance Council**[2] (**DSAC**), you will have access to a wealth of information, including contact information for the FBI special agent for your region so they can help during a crisis. If you or your organization is not part of these organizations, the **Internet Crime Complaint Center**[3] (**IC3**) is another option to choose, as they do allow businesses to file complaints of alleged cybercrimes.

Incident response software

You will need to lean on your experience, training, and tooling when an incident strikes. Your experience and training develop over time and by being exposed to various past incidents. Depending on your environment though, the methods will be the same but the toolls will vary.

Auditing and logging are used for two separate functions: operations and security. Operations teams use system logs to determine whether there are issues with an IT resource. These can include a service not restarting properly, a failed disk drive, or a network router not receiving routing updates. These logs are viewed under a different lens when it comes to security.

When reviewing logs for security, we may be interested in operational information as that could also pinpoint an issue; however, we are interested in many different things. These could be successful or unsuccessful authentication attempts. Has someone tried to perform a ping sweep against our network, or are we seeing malformed internet traffic, which could identify an active attack? What was the origin of the traffic coming into a web server? These should be logged on the IT resource and stored in a centralized location to correlate the data with other events happening on similar systems.

Users are hit with spam and phishing emails on a daily basis. Tools are available to help understand whether the email was malicious or not. With VirusTotal, you have the ability to check an IP address, URL, or hash value to see whether it is a known bad endpoint or file. With Maltego, you have the ability to perform searches against VirusTotal and other software right from the console, gathering information about the attack.

Osquery is a great tool for gathering IT resource information remotely. By using a common language, SQL, an engineer can access a server of any type and run commands to gather information. This information can include operating system level and type, IP address, the local host's file, and information about software on the system. This can come in handy when you quickly need to understand what happened on a server.

Simple Network Management Protocol (**SNMP**) is a free service that can be used on any networked IT resource. You can gather statistics on how a resource is performing, disk space, and memory usage. This information is needed to establish baselines for a system. Software packages such as Nagios can be used to query SNMP and store historical data in a database for review. If a system is out of the baseline, it can be configured to send email alerts to system engineers to review.

Many router and firewall manufacturers can send NetFlows to network collectors. A NetFlow depicts how an endpoint communicates with another endpoint across a network. It can display the ports used and the source and destination IP addresses. If you have an endpoint that is communicating with another endpoint that it should not be communicating with, NetFlows will detect this. Ntop[7], an open source NetFlow collector, is one such tool that can be used to collect NetFlow information from the network.

While you do not get full functionality, there are companies that offer both paid-for and free services for personal and commercial use. Cloudflare allows organizations to protect their websites using a **web application firewall** (**WAF**) to block attacks and log events. You can block your website from being attacked by Log4j or similar web-based attack methods. If you want to do this yourself, grab an Apache web server, install mod_security and mod_proxy, and configure the rules you require.

The preparation stage can be a lot of work to get set up properly, but it is worth it in the end. Having the plan on an off-site device or location so that it is easily accessible in the event of an outage is recommended. Now that you have prepared for an incident, it is now time to move forward with the detection and analysis phase.

Detection and analysis

Incidents come in all shapes and sizes, from the small ones, such as virus outbreaks or an employee experiencing a small breach such as an account takeover, to large ones, such as ransomware or **distributed denial of service attacks** (**DDoS**). You must train for the uncertain; however, you should focus your training on what will most likely impact your business. For some, it may be account takeovers, email phishing, and malware. For larger organizations, it could be website defacements or DDoS attacks. Some attacks, such as ransomware, every organization should be concerned with. Every organization will experience some type of attack, and it is a matter of what impacts the business the most that you should focus on.

A large piece of detection and analysis is documentation. You must document everything that has happened. This will not only help with your team's self-improvement but also in the event that you need to produce information to hand over to authorities. The next section will describe documentation in more detail.

Documentation

You will need to document everything that happened in your ISMS. This should include every incident, vulnerability, and threat directed toward your IT resources. Incident response is no different. To do this properly, you will need to create a template for an incident log to be filled out and kept in a document repository. Auditors, the C suite, and the board may ask you questions as to how you mitigated against a particular threat from your environment.

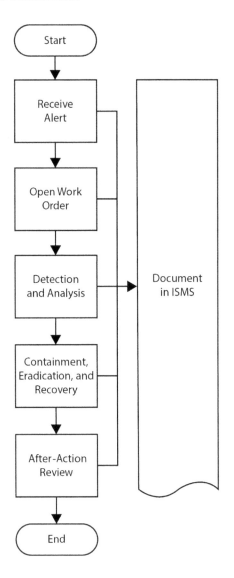

Figure 7.2 – Documentation in the incident response process

As soon as you get word that you have an incident, you begin working on the problem. The only issue is you are nose-deep into your computer screen concentrating and forget to do one task, creating a work order. This is the start of your documentation, as you are not sure what you are dealing with yet. As more and more users call the help desk, you are able to correlate the information to determine how widespread an issue it is. A problem ticket is then created to reference all the related incidents.

A timeline of actions must also be kept to understand what was done and when it was performed. This timeline will help in a number of different areas, but it will first tip off response times and who was involved. If you have a widespread incident and the first call came in at 8:00 am, and the first technician did not begin working on the problem until 8:15, that could indicate a response time issue. The timeline should also identify the people involved when a task was performed.

Documentation should include possible indicators of compromise, software applications used, hash values of payloads, information regarding the origin of the attack, and outside sources used. For instance, you come across an unusual file in the temp folder on a computer. You want to know what it is, so you generate a SHA-256 hash of the file and run it by VirusTotal or through Maltego to find out whether it is a malicious file. The timeline should capture the time and steps involved, but it should also include the forensic information too. When we define indicators, we are better prepared to understand what controls were put in place and why.

Defining indicators

Indicators of a possible attack can come from many different sources; however, there are really only two ways of detecting an incident, precursors and indicators:

- **Precursor**: A sign that an incident may occur in the future
- **Indicator**: A sign that an incident may have already occurred or that you may be experiencing the incident at that moment in time

While we tend to think of the incident in terms of indicators, how often are we proactive in identifying the precursors? Do you read cybersecurity news or pay attention to what is happening in your sector? Are you using **open source intelligence (OSINT)** to discover precursors in your environment? Do you watch social media for any information that may be interesting to the organization or scan the dark web hacker chatter?

I once ran a tabletop exercise for an organization where I made a statement that it was discovered on Twitter that a hacking group was making threats against organizations involved in labor disputes. This was not an indicator based on the definition. I wanted the team to think about the statement and make a judgment call on what to do next. Should they put in sensors to determine whether similar attacks were being made against the network? Could they gather intel from similar organizations as part of an ISAC or communication plan? Ultimately the SIRT came to the correct conclusion, but it was how the team got there that was the most interesting.

There are several different ways that you can write your own detection and response rules. For instance, YARA rules are used for the detection of malware or other potentially unwanted applications within an environment. Originally created by VirusTotal, YARA[4] is a cross-platform tool that can be used to scan systems looking for various types of viruses and other malware. It uses a C language format for rule creation, which allows you to search for Base64 to file hash values.

Playbooks, or runbooks, explained later in this chapter, provide step-by-step guidance on what an employee should do during an incident. Playbooks can also be used for how to respond to or contain an incident before, during, or after an event.

Containment, eradication, and recovery

The third phase of the incident response plan has three separate steps: containment, eradication, and recovery. At this phase, you are actively working on the incident, trying to find the source of the issue. Once the source has been identified, how can you prevent it from potentially spreading to the rest of the network? And after the incident has been contained, you need to remove the threat from the environment, and then recover the systems, and place the environment back in a normal working state.

There are plenty of unknowns when working on an incident. What are the indicators of compromise? Is this patient zero (the first infection point)? How many other systems have been affected? As you and the rest of your SIRT are working on the issue, you must find the source or the IT resources that are spreading the infection and prevent it from going elsewhere. As time is not on your side, you and the rest of the SIRT will need to rely on previous training and incident response playbooks. This is also the time to see how well your team responds to the incident and how the response can be improved.

Once the infection has been contained to single or multiple systems, it is time to remove the threat. This could be as simple as running a virus scan on the machine to something more challenging such as blocking a **denial-of-service (DoS)** attack upstream. It is also time to determine the initial source of the infection. How did the threat disrupt the IT resources on the network? Can that be prevented from happening in the future?

Preventing the threat may be another issue that you run into. Is there a patch that can be applied to the software? What is the procedure for contacting your internet service provider to get assistance in blocking a DoS attack? Was the vulnerability caused by a misconfiguration of a service? You will need to spend time determining the best course of action to prevent the attack from occurring again.

Lastly, you must recover from the incident. How do you put the pieces back together again to ensure that IT resources are working as intended? A disaster recovery, business continuity, or systems security plan will assist with this effort. A systems security plan should have all the necessary diagrams and information to help an engineer if they need to rebuild a system. Disaster recovery and business continuity planning will also help determine what systems should come up first and have their interdependencies identified to ensure proper recovery.

When focusing on containment and eradication, we must develop an interim and permanent corrective action plan. When creating these action plans, ensure they are documented for later review during the *after-action review* (discussed later in the chapter).

Interim corrective action plan

When you have either pointed out where the incident originated or you have a good idea, it is time to contain and eradicate the problem. An **interim corrective action** (**ICA**) plan is a corrective action plan in which you intend to put controls in to fend off the attack. These could be temporary compensating controls in which you plan to place permanent controls later or these turn into your permanent controls.

Suppose you are on the receiving end of a DDoS attack, and you call your internet service provider to put in a blackhole route[5] for your network. Or if you were hit by a ransomware attack, and you put in new password requirements or deny users from accessing the company's **virtual private network** (**VPN**). These are temporary solutions to prevent further attacks from happening. These scenarios however do not prevent permanent solutions from being put in place.

Make a running list of ICAs that were implemented; that way, once the incident is over, then you can either remove the controls or work on putting them in permanently. ICAs should be captured during the incident, and document where the control was implemented, who was responsible for its implementation, and how it was configured. This will also help guide the SIRT in placing permanent controls where they need to be, as you have an idea of how to contain and eradicate the incident.

Permanent corrective action plan

Now that you have identified where the attacker came from and how they got in, it is time to remove the weakness. Though **permanent corrective action** (**PCA**) plans are normally implemented either at the end of the incident or after the incident has concluded, these are controls that need to be in place to mitigate future similar threats. A PCA is normally captured during the lessons learned exercise, where team members discuss what happened during the incident.

PCAs can also consist of multiple small- to medium-sized projects to implement a control. This could mean that ICAs that were implemented during the incident stay in place until a permanent solution is identified and implemented. Risk registers are a common way of detailing the risk associated with the ICAs, and the permanent controls should be prioritized based on risk to the organization.

Basing the remediations on risk would be the best approach. For instance, you have a fairly large risk when it comes to scanning for vulnerabilities and patch management. While both are critical to the organization to identify risks, you have an ICA already in place to patch the identified vulnerabilities. Since you implemented the patch to mitigate the vulnerability, you still have a gap when it comes to identifying vulnerabilities. The PCA should identify both; however, now is the time to work on vulnerability scanning, as the ICA took care of patching the vulnerability.

In addition to patching one system within the environment, you should also identify whetherthe vulnerability resides elsewhere. If you eradicate the infection from one system, but others are still vulnerable, you will be spinning wheels trying to play catch up if another system becomes infected. This would be another PCA you will want to track to ensure the environment has been cleared of the threat and the weakness.

Post-incident activity

Now that you have contained and removed the threat from the IT resources and recovered from the incident, it is now time to bring the incident to a close. While this step may be overlooked, it is just as important to wrap everything up as it was when you were fighting the threat in the environment. This is your time to bring everyone together, debrief on the actions taken during the incident, and provide time to develop project plans for the items identified in the PCA.

After-action review

The incident is done, containment has taken place, and the threat has been eradicated from the network; this is the time to regroup as a team and review the actions taken. This is your time to debrief on what happened during the incident. An after-action review is your time to understand the course of action, the steps taken, and how the incident was brought to closure. Originally developed by the U.S. armed forces, this structured discussion is your feedback loop to improve your incident response capabilities.

We have all heard the term, "What happens in Vegas, stays in Vegas!"; well, the after-action review follows the same idea. An after-action review is done as a postmortem exercise in which the team comes together to discuss the actions taken during the incident. These discussions should be held using an open forum approach without fear of being reprimanded.

Discussions during an after-action review can sometimes get heated, and that is ok. What you are trying to accomplish is an understanding of what went great during the incident and what failed spectacularly. Failures can include not knowing the incident response plan, not knowing where the documents reside, and not truly knowing the technical environment. These failures identify weaknesses in your incident response program. These findings, and the steps to overcome the failures, should be documented. The after-action review is your feedback loop of information gathering, which leads to continuous improvement.

An after-action review should take place no later than 2 weeks after the incident closed. This will allow ample time to gather your thoughts on what went right, what went wrong, and your ideas for how to improve the process. The review should include all the members of the SIRT and anyone else who may have been part of the incident response. Questions should at least include the following:

- What individuals or groups were part of the incident?
- What was the timeline of events?
- Were documented procedures followed?
- Was information provided in a timely manner?

- What things could assist when responding to future events?

- Is there anything that has not already been asked?

These questions are meant to spark conversation as to what happened and why. The question, "Were documented procedures followed?" is meant to identify whether or not a procedure for that event exists and if it does, was it followed? If it does, and no one followed it, then there is a problem with training the SIRT on the procedure. If the procedure does not exist, it is time to create one and evangelize it to the group.

For situations when time is of the essence, did the SIRT respond quickly enough? Were people talking during the incident, providing much-needed information to others? Was there a war room where everyone could discuss their actions? All this then leads to what would make responding to an incident faster. What would be the biggest improvement for the team?

The last question is meant to be left wide open for comment. Anyone can have a voice during this discussion as it is intended to allow everyone to speak up. The team should learn how to vocalize their fear, uncertainty, and doubt about how the SIRT responded poorly along with praises for what went right. The entire SIRT learns a lot from the last question, as it provides constructive criticism to the entire group.

Follow up with law enforcement

If you do not have the resources to perform forensics or need assistance when an incident has occurred, leaning on law enforcement is your best bet. They are there to assist during a cybersecurity event. However, much like your own resources, they too can be strapped for time and the ability to help.

You will need to assess your local, county, and state law enforcement capabilities to help during or after a cyber incident. Local law enforcement may not have the background or experience in cyber to fully assist with an incident. If you live in large cities, those resources may be available; if not, I suggest building a relationship with larger organizations, such as state police or even the FBI. There are plenty of resources that you can use at your disposal when contacting federal law enforcement agencies. The IC3 has a plethora of resources that you can use, including filing complaints. It would also be recommended to register with the InfraGard program. This program is a partnership between the FBI and the private and public sectors to provide cybersecurity information and assistance when needed.

Building the **incident response** (**IR**) plan is a necessity when you and your team need to respond to a given incident. Building relationships with internal and external organizations is also key. You cannot do this alone; it takes a village. You, at some point, will need to lean on others to build out containments and eradication of incidents within your environment. The next section will help build out those response plans and assist with the various steps when reacting to a given incident.

Incident response playbooks

When you are in the heat of an incident, every minute is valuable, and making the most of them can only be done through quick responses. Incident response playbooks are your manual for responding to a given incident. The playbooks provide step-by-step instructions for what to do when an incident does occur. Playbooks have to be detailed enough so that a junior administrator or a non-technical person can follow the steps outlined to perform an action.

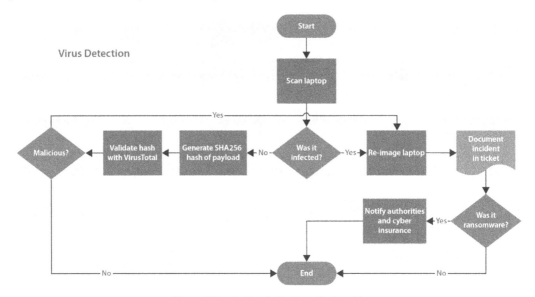

Figure 7.3 – A virus infection playbookk

The preceding diagram depicts an incident response playbook for how an organization could respond to a virus infection. We begin by scanning the laptop with the antivirus software installed on the machine. Then a decision is made, was it or was it not infected? That decision then spawns two separate directions for a course of action. If it was infected, re-image the laptop, document, and move on, whereas the other decision validates that the payload was not malicious using VirusTotal[6] and closes out the ticket.

While that was a high-level discussion of how an organization could respond to a virus outbreak, it should be detailed and easy to follow. Improvements to the incident response playbooks should be made on a regular basis. Weaknesses discovered in the playbook should be called out and reviewed at an after-action review. Playbooks, much like policies, standards, and procedures, should be reviewed at least every 2 years. However, it is recommended that they be reviewed after an incident to ensure that you are receiving the results you intended.

Train like we fight

The U.S. military has a saying, "train like we fight, fight like we train." Athletes train all year to be the best at their game, so why would training for an incident be any different? This reminds me of a scene from Apple TV+'s series, *Ted Lasso*, "We're talking about practice!" It takes roughly 10,000 hours of practice to become a master at something. That is roughly 13.5 months of repeating the same thing over and over again until you begin to develop muscle memory.

There are a few different ways to practice or train your SIRT to develop that muscle memory, including tabletop and live-action exercises. Tabletop exercises are a low-cost/no-cost solution for training your employees, whereas live action is a simulation of an attack. The following are only a couple of the ways you can get your team up to speed on how to respond to a cybersecurity incident.

While this is not an extensive list of incident response tests, the three depicted next are the quickest low-cost/no-cost methods that you can do initially. While other tests, such as those that introduce disruption into the environment, are great to run, they are not ones that you can do on a regular basis. They also require significant planning due to potential downtime being introduced into the environment.

Walk-through exercises

Walk-through exercises are the quickest way to get an understanding of your incident preparedness. You will need to have a checklist of objectives that you want to test to ensure that they are configured correctly. This may also require pre-planning if there are data center walk-throughs, as many do not let just anyone in.

A walk-through exercise is one where someone on the SIRT goes through a checklist for how the team should react to a given situation. Did an alert come up on the screen? How would you and your team react to that alert? Did the team follow the process? How about a walk-through of someone implementing a firewall rule to stop the spread of a virus outbreak?

A walk-through is the first step in developing muscle memory for how the team should respond. This is like a guided tour, walking through each step of the process to better understand how the team should react to an incident.

Tabletop exercises

Think of a tabletop exercise as an analog way of training. No computers are needed for this as these are scenario based. The SIRT will be handed one question at a time and told to work through the problem. The problems given do not necessarily have to build upon each other; however, I have seen plenty of successes either way.

You select a scenario, say ransomware. From there, develop a list of scenario-based questions and tailor it to your environment to make it meaningful. Provide the SIRT with a question and allow the team to create a solution that will work for that scenario. These are typically time-based questions to encourage the team to work quickly and efficiently.

At the top of the given time, you give them another problem to work on and build upon each other. In the example of ransomware, you present a problem that someone is calling the help desk because they are no longer able to open files on their desktop. Then it spreads to the file shares and infects the servers, and now no one can log into a server because it has taken over the LDAP servers.

Live action exercises

Live-action exercises can come in several different methods; however, this requires access to a computer. For example, with red team/blue team exercises, you have two teams trying to break into networks while the other team is fending off the attacks. These types of scenarios can also be expensive to put on; however, they teach the SIRT how to react to a particular scenario.

The red team is looking for holes in the IT resources. Once a vulnerability is discovered, the team will attack it to gain access to the other team's network. Launching attacks on the website or performing a SQL injection are just some of the ways that the red team can attack. Tools such as Metasploit and Kali Linux are typically the ones you would see being used. These tools are free to use; however, there is a learning curve when trying to use them. Training and experience are the only ways someone will feel comfortable.

Live-action scenarios can be expensive, as you should not do this in a production network. Too many things could go wrong in a live-action scenario, as attacks are actively being launched against the other side. The blue team could put in a countermeasure and accidentally affect production systems.

Cyber ranges are also a way to perform live-action incident response training in an enclosed environment. Cyber ranges are typically developed by third parties and offer their services to customers. The nice thing about a cyber range is that the tools are usually provided for you, and you do not have to worry about affecting your production network. The scenarios are held in an isolated environment, sometimes enclosed in a *sandbox* where the attacks do not accidentally go across the network or the internet.

Summary

I am not one for puns or clichés, but the saying, "It is not a matter of if, but when…" should be ingrained into everyone's brain. Whether the attack is large or small, you and your company will experience some type of attack. It is how you prepare for it that will dictate whether the attack is successful or not.

In this chapter, you have learned about the NIST incident response methodology and it's four main stages which you can leverage to prepare your team for when an incident occurs. As mentioned, develop incident response playbooks to help yourself and your team sorespond to an incident. Document, document, document…and if you feel that you have not documented enough, document some more.

Once an incident has been identified, you will want to contain it so that it does not spread to other parts of the infrastructure. Once it has been contained, remove it from the environment and recover any service disruptions that may have been caused.

We then learned about the importance of developing corrective action plans to prevent or reduce the likelihood of the incident from reappearing. This may include the installation of software patches, opening or closing firewall ports, or replacing a system altogether. The idea is to get your IT resources back to running at expected baselines and close off any vulnerabilities.

Lastly, remember to bring everyone together at least 2 weeks post-incident for an after-action review. This session is meant as a way to express what went right and what went wrong during the incident. This is meant as a way to get to the root cause of how the incident occurred and how the team responded to the incident.

In the next chapter, we will focus on cybersecurity awareness and training and how that will impact not only your team but the rest of the organization.

References

1. FBI InfraGard: `https://www.infragard.org/`

2. Domestic Security Alliance Council: `https://www.dsac.gov/`

3. Internet Crime Complaint Center: `https://www.ic3.gov/`

4. YARA website: `https://virustotal.github.io/yara/`

5. Cloudflare blog post on DDoS blackholing: `https://www.cloudflare.com/learning/ddos/glossary/ddos-blackhole-routing/`

6. VirusTotal virus scanner: `https://virustotal.com`

7. Open source NetFlow tools: `https://www.ntop.org`

8

Security Awareness and Training

The cybersecurity landscape continues to evolve, almost by the minute. There are always new threats, new vulnerabilities, and new ways of breaking into an application. Sure, there are bug bounty programs that reward security researchers for their time in disclosing a vulnerability; however, it takes time to patch, test, and release them to the masses. There is a time delay between discovering a vulnerability and fixing it. How would you know what to do to prevent your organization from being threatened by such a vulnerability?

This is one of the reasons why cybersecurity awareness, training, and education programs are so important for not only your security team but also IT and the rest of the organization. As new threats are discovered, there needs to be a way for you, as head of security, to promote that information. This only comes through releasing awareness videos, banners, and other documentation.

You may think that cybersecurity is a technology problem; it is not. It is a people problem, and employees are your first line of defense. They are your eyes and ears for anything and everything cybersecurity-related. Because of this, it is important to get information out to your users as quickly and up-to-date as possible.

Training, awareness, and education are not synonymous; do not conflate them. Whereas awareness is focused on high-level information and short-lived 5–30-minute videos explaining what to do and what not to do, training is more in-depth. Education focuses on classroom-based teaching in a structured setting, whereas training focuses on a single subject and is more of a deep dive than just high-level concepts. This difference is precisely what this chapter will try to firmly establish.

In a nutshell, this chapter will focus on the following:

- Understanding security awareness, training, and education
- Setting up a security training program
- Continuous improvement
- Training for compliance

Understanding security awareness, training, and education

"What if we train them and they leave? What if we don't and they stay?"

I have come across that quote once or twice during my tenure as a cybersecurity professional. It holds true in so many different aspects of the IT and information security fields. Employers can be reluctant when looking to train professionals, as competition is fierce from both a company and employee perspective.

I have worked for a few organizations that have had dedicated training budgets for their employees. This was a nice perk of the job. The company not only understood the necessity for continual education but also the power of networking with others in the field. It showed that the organization was serious about the program being developed, or in use, but was also serious about ensuring that its employees were well informed. They were informed to make the best, educated decision with the information they had at hand.

I have also worked for organizations that do not have a dedicated budget for training. These organizations either expect an employee to already have the knowledge that pertains to their job or have ways of gaining an understanding through the job training. While these organizations are not necessarily bad, they do present a dichotomy. How do you expect your organization to grow and stay ahead of technology or an adversary if you do not provide the tools necessary for the employee?

While some organizations that do not have a budget offer other forms of employee training, some do nothing at all. They expect the employee to know or learn how to do a particular job on their own time. This can be detrimental to the employee, as they may never have been given the tools to succeed in performing their job function. There is also another famous quote that says, *"If you think education is expensive, try ignorance."* Ignorance will be more expensive due to poor architectural design, implementation, and the reality of having to redo a project more than once.

One of the many aspects that draw so many people to the cybersecurity field is that it is ever-changing. Something that you learned a decade or even a year ago could be obsolete today. For instance, Moore's law states that the number of transistors on a microchip will double every 24 months. However, it has been stated that Moore's law has since been broken, with the number now doubling every 14 months[1].

Cyberattacks are no different. The **Open Web Application Security Project (OWASP)** publishes the OWASP Top 10, a list of the top 10 web application vulnerabilities. First published in 2003, it is the leading standard that organizations use to protect against web vulnerabilities. If you do not allow your employees to gain the additional knowledge or training to fend off these types of attacks, then you will become vulnerable.

This approach does not solely pertain to cybersecurity; however, it should be part of your cybersecurity program. Again, it takes 10,000 hours to become a master of something. It will take you a little over a year and a month of continual 24x7 work, with no breaks, to develop the necessary muscle memory. There is also a saying that goes, "If I do a job in 30 minutes, it is because I spent 10 years learning how to do that in 30 minutes. You owe me for the years, not the minutes." This holds true for those that have a background or the knowledge to do a job efficiently and effectively.

As mentioned earlier, while awareness, training, and education may seem synonymous, they are not. Each one of these has a distinct focus when it comes to consuming information. The following sections will walk you through each.

Awareness

Awareness is focused on how an employee, or the person learning the material, receives information from a teacher. The information is passively consumed and can be provided through **computer-based training (CBT)**, banners, flyers, and audio or video. Awareness is about getting information to a user regarding a specific topic on a regular basis.

Think of awareness as a means to push out content to the user base that can be consumed through direct or indirect means. Perhaps put banners up on walls telling your users not to click on suspicious emails. The screensaver on your computer could show slides of cybersecurity-related topics, such as devising good passwords and multifactor authentication. Awareness topics should be high-level and tailored to anyone, whether they are technology-savvy or not.

When planning to roll out your cybersecurity awareness program, make sure to align it with a common theme or framework. Pick topics that will have the greatest effect on your users. For instance, social engineering is a common theme used by attackers to gain access to your network. Picking awareness themes such as phishing, spam, vishing, and other common social engineering attack vectors will have the greatest impact on your business.

There are vendors that specialize in providing cybersecurity awareness courses for you and your employees. Ensure that the videos are not too long that you may lose your audience but just long enough to get the point across. Videos should be anywhere from 5–20 minutes and can be broken into different themes within the same topic. Many offer computer-based awareness videos that touch on a number of different topics, including the following:

- Email protection
- The dangers of connecting to untrusted networks
- A clean desk policy
- Acceptable use
- Tailgating

Many cybersecurity awareness vendors offer quizzes or tests, either during the video or at the end. This is a great way to discover how effective the cybersecurity awareness program is and where any gaps may be. They also provide metrics on how many have seen educational material, if they started to look at it and stopped, or have not seen the material at all. These are all key areas to review to ensure that everyone is viewing the material.

This is an area of exponential growth too. It is estimated that the cybersecurity awareness and training market will exceed $10 billion annually by 2027 (Braue)[2]. You may get inundated by all types of attacks throughout the day, month, and year at your company, so you must remember that people are the first line of defense in the program. They are equally important to the other controls that you implement.

Training

Though you are still learning a particular topic or theme, training differs from awareness. Training is focused on a single topic and can provide high- and low-level overviews. Cybersecurity training can span a wide gamut of topics, from administrative to technical controls, and topics such as performing assessments, implementing governance, and viewing organization risk. Other topics include vendor-specific firewall administration, which can span beginner, professional, and expert levels.

While cybersecurity awareness topics should be given at least every quarter, training is normally provided once a year and typically lasts a week. For an employee to obtain training, an individual or group would have to travel to the trainer's facility. This is typically the best course of action, as their facility will have all the necessary equipment to perform the training.

Training is typically more expensive than awareness programs too, ranging anywhere from $1,500 to $10,000 in some instances. Additional costs must be factored in too, such as travel, food, and hotel rooms. A good guesstimate of the total cost for a week's worth of training is close to around $5,000–$6,000.

Bootcamps are typically more expensive than traditional training programs, as they prepare a student to pass an exam. Some organizations, even the U.S. **Department of Defense (DoD)**, require certain certifications for a job role. For example, the DoD's 8570 requires a certification to be obtained out of the list provided[3].

Certification testing differs between vendors as well. Many are used to the true/false, multiple-choice questions of traditional tests. Some certifications will require a student to perform simulations on how to configure a piece of equipment or ask the student to troubleshoot why a configuration is not working. My favorite certification tests are the ones that place the student in front of a live system to perform setup and troubleshooting tasks. There are a hundred different ways to implement something, which allow students to do it their way and not, say, the Microsoft way.

Considering that training can be a time-consuming endeavor, it may also make sense to train en masse or an entire group of employees within a particular department. While some organizations may train one or two individuals on a particular subject or technology, other organizations may train everyone in a department. This all aligns with business objectives and how you want to maintain a certain level of knowledge of a particular product.

There have been times that I have seen individuals from different departments get trained due to their job responsibilities. For example, the implementation of a new firewall could involve someone from networking, security, and compliance. In addition to having employees from various departments, those departments also trained their analysts and engineers to support and maintain a particular technology.

Education

I was never much for school growing up, but we all had to do it. We have all had different experiences in the education system – some good, some bad – but we are all better off because of it. Education is a formalized, multi-disciplinary approach to teaching necessary skills to students. Not only does it provide the necessary knowledge for teaching the skills needed for the particular field you are looking into but also provides a well-rounded education in other topics as well.

Education is the most expensive way to gain the necessary knowledge to enter any market, not just cybersecurity. Whether itis your tax dollars paying for the public education system or putting yourself (or maybe even a kid or two) through college, it can be costly. According to U.S. News, the average annual cost of tuition at a public university was $10,388 for the 2021–2022 school year, and it was $38,185 at private universities (Powell)[4]. That plunges you into debt before you even receive your first real paycheck.

Employers seek students that have gone through the education system in order to achieve a degree in a certain field. Many, but not all, require that the student have at least an associate's or bachelor's degree prior to having an interview with the company. However, obtaining a degree, on average, will provide you with the highest return from income generation.

Some employers will hire candidates without a degree and take years of experience in lieu of them having a formal degree. Organizations may put stipulations on the candidate's background and education before extending an offer. For instance, many job descriptions will state that they will take 8 years of professional experience or a 4-year degree from an accredited college or university. Elon Musk, founder and CEO of Tesla, says he has no use for college, arguing that there are plenty of billionaires that dropped out of college to start their own business or never went in the first place.

Whether it is cybersecurity awareness material, training, or education, it is important to make sure that your employees have the latest information to protect themselves and their organization. The employees at your organization are your first line of defense and should be provided with material so they too can understand what to do when faced with several different attacks.

To make a successful program, you will need not only executive backing but also to create an environment for continual education. This can be accomplished through the preparation and development of an internal security program, so let's get right into it.

Setting up a security training program

To create a continual learning program, you first must receive executive backing to do so. Remember, the business is paying the bill for your employees, not you. Some organizations may not take kindly to you expecting employees to dedicate a certain amount of time for continual education on their dime. This is why you must develop the business case, obtain their backing, and build metrics around why training and education not only benefits the employee but the business too.

Establishing the need for a security training program

To get your cybersecurity awareness and training programs off the ground, you will have to first establish a need. You will need to understand what deficiencies you have and then figure out how to overcome them. This does not mean that you or your employees do not know or understand their job role; it is a matter of figuring out what the weak points are and the type of training that is needed.

If you have a big upcoming project where you are swapping out old firewalls for new ones, it is entirely possible that your security or firewall administrator will need to understand how to operate the new equipment. The same goes for any type of technology. The quickest and best way to bring someone up to speed is through a training class.

Obtaining executive support

Much like with anything in your cybersecurity program, you will have to get executive backing for your awareness and training program. The company is the one paying the bill, so it is ultimately up to them to send someone to a training class or seminar. If your company has a training budget, this may be an easy request. If the company does not have a training budget, maybe it is time to start one.

You will need to begin with your management team. They are the ones that you will have to win over to spend money on awareness and training. Have your elevator pitch ready so that you can talk about the necessary points that you want to get across. Also, discuss the need for training. Is it for an upcoming project? Is something lacking in a person's skill set? This is where the information gathered while establishing the need will come into play.

Developing metrics

You should also develop and keep metrics for your awareness and training program. Metrics will help you keep track of who went to training, the type of training they received, and the possibility to help others in a similar scenario. Metrics can also help with obtaining certifications as well.

Keeping track of the types of training you and your employees have taken is a necessity that will help understand the courses taken by your employees, whether it was beneficial or not, and whether you should send others to the same type of training. It is also helpful to know the training facility that will be hosting the class and whether they are meeting your expectations.

Some organizations want to know and keep track of the various certifications their employees have or need to obtain. They sometimes do this for marketing reasons, so they can say, "*Hey, look at how many CISSPs we have on staff!*" or "*Tony is our senior engineer and is a triple CCIE!*" This also qualifies the company to bid on certain jobs or contracts. Would you pick an organization to perform an assessment with someone who does or does not have a **Certified Information Systems Auditor** (**CISA**) certification? What about the implementation of a new Cisco environment from someone who has Juniper certifications? Consultants and their employers want to ensure their team is trained on the latest technology, and certifications show that they have the knowledge to perform a job.

Examining course objectives

Not all training courses are alike. However, they most likely have one thing in common, which is that a course must follow a list of objectives. Course objectives are like a **build of materials** (**BoM**) or a **statement of work** (**SoW**). They will tell you exactly what they will cover during the course so you can make better decisions on what training you want to take.

Most vendor-specific training must be taught by an instructor that has already obtained the certification, or one higher than the one they are teaching, and received a score at a certain percentage. This means the instructor has the background and knowledge to present the material appropriately. This also means the instructor is capable of answering questions regarding the subject matter. For example, an instructor for a **Cisco Certified Network Associate** (**CCNA**) course must have achieved a CCNA or higher-level certification and have obtained a certain score on their exam. This ensures the instructor has the background and knowledge to not only teach the course but to also answer any question a student may have during it.

Vendor-specific courses also come with their own standard training materials. This means that if you have a distributed workforce and you need to send employees to different locations, they will receive the exact same materials. This does not mean, however, that the equipment will be the same at each location. As equipment may be different, performing an action one way does not necessarily mean that it will work the same way on all equipment.

Continuous improvement

What can you do to keep yourself updated on the latest news and trends in cybersecurity? This is a question I get asked a lot, and in cybersecurity, it may seem like you are continuously drinking from a firehose, but that is what the field is about, right? As new attempts to compromise a system are developed, attack vectors, cybersecurity research, and new ways of trying to uncover flaws within a system happen regularly.

Really Simple Syndication (**RSS**) is one way to gather cybersecurity news and place it all in one location. News outlets from across the world write stories about cybersecurity and publish them for free online. News outlets are not the only ones that produce cybersecurity news for everyone to read; cybersecurity bloggers also create content for their readers. Famous bloggers and security researchers such as Brian Krebs and Bruce Schneier write articles or report interesting cybersecurity news stories on their websites.

Low-cost and no-cost computer-based training is also available. Websites such as Udemy[5] allow content creators to develop cybersecurity training material and resell it to customers on the platform. There is a ton of content on the Udemy site, such as cybersecurity for beginners and more advanced courses, ethical hacking, risk management, and privacy. They also have content on application development and scripting to help with automating your infrastructure.

Publishers also allow you to read their content online for a fee. For about the price of a book, you can subscribe to a publisher and read their entire catalog monthly. This can sometimes be a cheaper route if you are a person like me who tends to read a lot of technical books about computers and cybersecurity.

Many organizations offer tuition reimbursement, encouraging their employees to go to school and earn a degree in the field they are working in. This greatly benefits both the employee and the company. As the employee learns new concepts, they can immediately apply them to the job they are performing. Some coursework allows for hands-on training, providing people the ability to work with equipment in a lab setting. This reinforces the concepts learned in the classroom by applying them to physical equipment in a non-production environment.

Most higher education or trade schools do not necessarily offer the right courses for you to meet certain obligations. For instance, training for certain compliance regulations require you to go to a special trades academy to gain that type of knowledge. Remember, security is a mindset, not something you set and forget about. Security training is much the same and many regulatory bodies expect that corporations provide or ensure that their employees maintain IT and security training.

Training for compliance

Cybersecurity awareness training is also a necessity for compliance reasons. Regulatory bodies want to ensure that those organizations that are performing a high-risk task can do so securely. In addition to that, they also want to ensure that those who handle the data know and understand the risk they are taking.

Requirement 12.6 of the **Payment Card Industry Data Security Standard** (**PCI DSS**) requires organizations to implement a cybersecurity awareness program. The PCI council made this a requirement years ago due to a lack of training and understanding of compliance. Some of the questions asked by the requirement are as follows:

- Do new hires receive cybersecurity awareness training?
- Do employees receive cybersecurity awareness training at least annually?
- Are employees aware of the importance of cardholder data?

The **Cybersecurity Maturity Model Certification** (**CMMC**) also has cybersecurity awareness and training requirements. Section 3.2 of NIST SP 800-171 requires organizations that store, process, or transmit **controlled unclassified information** (**CUI**) to perform cybersecurity awareness training for their employees. CMMC also wants to ensure that those who are working with CUI data have been formally trained to securely perform their job functions. This again is different from cybersecurity awareness, as they are looking for a formalized training session on securing data.

ISO 27001 is yet another example of where a cybersecurity standard requires an organization to have a cybersecurity program. Section 7.2.2 of ISO 27001:2013 requires all employees and contractors to not only receive cybersecurity awareness training but also receive formalized training. ISO takes this a step further and stipulates that employees and contractors shall be updated on new or updated policies, standards, and procedures.

These are just a few of the standard requirements that force organizations to build, support, and maintain a healthy cybersecurity awareness program. Regulatory bodies are starting to understand that cybersecurity is not just a technology problem; it is a people problem too. IT and security staff are not the only ones that are responsible for maintaining your cybersecurity program; it is a team sport and must include everyone in the organization.

Summary

As the cybersecurity landscape continuously evolves, your employees will need to adapt – quickly. Cybersecurity awareness is all about reinforcement for all your employees. It is recommended that awareness videos, posters, and flyers are pushed out on a monthly basis. As you have learned in this chapter, training involves more of a deep-dive approach to a particular technology or subject matter, while education is a prolonged approach to studying and mastering a subject matter.

Instead of your employees being subjected to some type of continual improvement of cybersecurity subjects, they can be trained as a compliance requirement. Many of the leading security frameworks and regulations require that staff members, especially those who are a part of IT or perform some type of data handling, be certified in their field or have a certain type of qualification. If you fail to provide training to these individuals, it could mean the end of your compliance.

Remember, as the head of security, you must establish a need for training your employees. Build a business case for how you would like to train your staff and with whom you would like to attend certain courses. Courses range in price and delivery, so you must ensure that a course is right for your employees. Encourage your staff to continue to educate themselves as well. Set aside time in their day to allow for self-improvement and self-study. The organization must have your backing on this too, or your efforts will quickly be directed toward a different topic or need. Continual education is important for both an employee and a business.

As we conclude *Chapter 8*, we look forward to network security in the next chapter.

References

1. *AI Machines Have Beaten Moore's Law Over The Last Decade, Say Computer Scientists*: `https://www.discovermagazine.com/technology/ai-machines-have-beaten-moores-law-over-the-last-decade-say-computer`

2. Braue, David. *Security Awareness Training Market To Hit $10 Billion Annually By 2027*: `https://cybersecurityventures.com/security-awareness-training-market-to-hit-10-billion-annually-by-2027/`.

3. *DoD Approved 8570 Baseline Certifications*: `https://public.cyber.mil/cw/cwmp/dod-approved-8570-baseline-certifications/`.

4. Powell, Farran. *See the average college tuition in 2021-2022*: `https://www.yahoo.com/entertainment/see-average-college-tuition-2021-130000169.html`

5. Udemy online courses: `https://udemy.com`.

Part 3 –
Technical Controls

When we think of putting controls or safeguards in place, we typically think of using technology to compensate for a deficiency. IT resources can restrict traffic from accessing a resource, prevent authentication due to wrong credentials, and fend off attacks which come from multiple sources. Yet with all these technical controls in place, organizations still fall victim to cyberattacks.

Organizations may put in place safeguards to fend off attackers searching for open ports. They place antivirus software on servers and end-user devices to search for malicious software that has been placed on a machine. Additionally, IT administrators offload logging capabilities to a SIEM or log correlation engine to centralize log management. Businesses also purchase DDoS prevention services to intercept attacks from hundreds if not thousands of devices on the internet.

While we tend to use network resources to mitigate incoming attacks, we must also heighten our response to attacks against the software we use every day. So, *Part 3* of this book will discuss the basics of technical security controls and explore new ways of protecting your IT resources from attacks.

9
Network Security

The Internet is a series of tubes. —Senator Ted Stevens

When you think about it, Stevens is about half right, but the internet is so much more. The internet is Wi-Fi, fiber, e-commerce, banking, stock trading, social media, video calls, and now blockchains and the metaverse. The introduction of blockchain technology and Bitcoin brought about Web 3.0. This term was coined in 2014 by Gavin Wood, one of the founders of the Ethereum Foundation. Web 3.0 is very much alive today (at least at the time of writing), though it seems like we are at another transition point. Are we seeing the dawn of Web 4.0?

30 billion devices all connected to a medium that allows you to communicate with the other 30 billion devices on the internet causes some concern. How do you protect your devices from the rest of the internet? How can you recognize what devices are connected to your network? Are those devices even authorized to be on the internet? How would you protect an **Internet of Things (IoT)** device that has very limited hardware or software capabilities? The old saying "How do you protect a computer? Seal it in concrete and drop it in the ocean" really does not work for business.

In this chapter, we will discuss some of the fundamentals of networking and network security. First, we will review the **open systems interconnection (OSI)** model and IPv4 subnetting, and then look at the various types of network security, including firewalls, **intrusion detection systems/intrusion prevention systems (IDSs/IPSs)**, **data loss prevention (DLP)**, virtual private networks, and some new concepts related to cybersecurity, including zero trust and micro-segmentation.

All in all, the following topics will be covered in this chapter:

- The history of the internet
- The OSI model
- IPv4 addressing and micro-segmentation
- Traditional network security
- Networks of today
- Building trust into the network

- Virtual private networking and remote access

- Getting to know the "zero trust" concept

- Understanding firewall functionality

The history of the internet

The creation of the internet has two distinct storylines. The first message that was sent across the internet was done by students from UCLA and Stanford, but the internet also had a different reason for being developed. During the 1960s, at the height of the Cold War, the **U.S.** government needed a way to send electronic messages to other government entities to make sure they got to their destination. Why? Because they wanted to ensure that in the event of a nuclear war, messages in an electronic format could still reach their destination.

This brought about the **Transmission Control Protocol/Internet Protocol** (**TCP/IP**) and its ability to be rerouted across a decentralized node network which helped the government with its objectives. With TCP/IP, you can send a message, in chunks, have it routed across different network links, and receive the message in its entirety at the end. So, if one **point of presence** (**POP**) is offline, it could reroute the rest of the messages across a different POP and still make it to the destination – that is, if the destination is also online and each POP has multiple egress links.

On the other side, computer scientists from UCLA and Stanford were set to make history. They wanted to send a message from one computer to another; however, it did not work as they had hoped. As the two computer scientists sent a message stating LOGIN to the other side, the computer crashed, sending only LO. While this appeared to have failed, it was a huge success in that this was the first time in history that a message was sent between two distant computers. This was the beginning of the **Advanced Research Projects Agency Network** (**ARPANET**).

According to Help Net Security, there were roughly 22 billion devices connected to the internet in 2018. In 2022, the expected number of devices grew close to 30 billion, and this number continues to grow. As we begin to look to the future, it seems like everything will have a connection to the internet. IoT devices continue to push that number even higher, with everything from toothbrushes, streetlights, and heart monitors to your car eventually having a connection (if it doesn't already).

As you can imagine, and as iterated earlier, 30 billion devices all connected to a medium that allows you to communicate with the other 30 billion devices can give rise to security concerns. There are quite a few different ways to safeguard your corporate network from a number of different internet-based attacks. The OSI model, which we will look at next, will come in handy when discussing firewall architectures and protections.

The OSI model

Have you ever heard someone say, "That is a layer 1 problem." Or maybe you have implemented a layer 7 firewall or a layer 3 switch at one time in your career. These situations relate to the OSI model. This model has been around since 1984 and is a fundamental part of understanding networking.

The OSI model's structure is as follows:

Layer Number	Layer Name
7	Application
6	Presentation
5	Session
4	Transport
3	Network
2	Data link
1	Physical

Table 9.1 – OSI model

Let's look at each of these layers more closely:

- **Layer 7: Application**

   While the application layer is usually associated with actual applications, it is meant to represent the protocols used. The application layer is what allows an end user to interact with the application. Protocols used at the application layer include FTP, SMTP, and HTTP.

- **Layer 6: Presentation**

   Have you ever used hard drive encryption, possibly FileVault or Bitlocker? Or maybe you've used VeraCrypt for file-level encryption. The presentation layer is much more than just encryption software. This layer is meant to translate a file, for instance, into something that the application layer can use and present to the user.

- **Layer 5: Session**

   When two computers communicate with each other, they have to open a session. The session layer is responsible for establishing communication first by opening and then closing that communication. It is also responsible for keeping track of the data that is sent across the internet. Without keeping track of the data being transferred, it would have to start the communication over again from scratch.

- **Layer 4: Transport**

  While the session layer is responsible for opening and closing the communication channel between two computers, the transport layer is responsible for data transfer. The transport layer breaks the data into chunks before sending it to the network layer.

- **Layer 3: Network**

  The network layer can send traffic between two different networks. Devices designed for the network layer have routing tables, which are defined maps of networks the device knows so that it can send traffic back and forth. Devices that do not know the destination network will send packets to an **internet service provider** (**ISP**) and allow the ISP to send the traffic on your device's behalf.

- **Layer 2: Data link**

  Two computers that want to communicate with each other on the same network do so using the data link layer. The data link layer is used to send frames between two computers. When a computer communicates with another computer on the same network, it uses a burned-in address or **media access control** (**MAC**) address.

- **Layer 1: Physical**

  The physical layer deals with the equipment that transfers data to and from a physical endpoint. This includes network cables, both copper and fiber, and network hubs. Devices at the physical layer communicate using bits, which is done in the form of 0s and 1s.

Each layer of the OSI model communicates with the layers directly above and below it. For instance, the network layer can only communicate with the transport and data link layers. The physical layer cannot talk directly with the presentation layer, for example. The OSI model will come in handy when we start talking about network equipment.

The first three OSI layers

Traditionally, network hubs, switches, and routers communicated at layers 1, 2, and 3, respectively, so we will focus our attention on these three layers specifically. Network hubs were prevalent throughout the 1980s, 1990s, and early 2000s. While rare, hubs are still in use today. Network hubs brought two or more computers together so that they could communicate with each other on the same network. The difference between a network hub and a switch is that a hub forwards all traffic through all hub ports. This means that if a computer were connected to five other computers on the hub, when that computer communicates with another device, the hub replicates the same network traffic, and the same communication is sent out the other five ports.

Getting physical with layer 1

Network hubs were a big upgrade; however, they posed a security problem. Not only would the hub replicate the same traffic and send it to other devices connected to the hubs' ports, it also made sniffing network traffic simplistic. Someone could essentially load up a packet-capturing device and grab all the traffic across the wire. All the computers were also in the same collision domain, which meant that when one computer began sending traffic to another device, no other computers were allowed to. If two computers communicated at the same time, a collision would occur and each computer that tried to talk would back off and wait its turn again. It was not until the introduction of the network switch that this changed.

MAC addresses and layer 2

The introduction of the network switch was a game changer for the networking field. Not only did it reduce the collision domain down to a single port, but it also kept track of MAC addresses communicating on that port. Network switches keep track of which devices are connected to a port through a MAC address table. The following is an example of a switch's MAC address table:

```
Mac Address Table
--------------------------------------------

Vlan    Mac Address        Type            Ports
-------    --------------------    -------------    -----------
        1            aaaa.aaaa.aaaa    DYNAMIC      Gi1/0/1
        1            aaaa.aaaa.aaa1    DYNAMIC      Gi1/0/2
        1            aaaa.bbbb.cccc    DYNAMIC      Gi1/0/3
```

This is a very simplistic view of a MAC address table from a network switch. Previously, a hub would not keep track of this type of information; however, a switch will. When the device with a MAC address of `aaaa.aaaa.aaaa` wants to speak to `aaaa.bbbb.cccc`, the switch will know that traffic coming in from port `Gi1/0/1` will be destined for port `Gi1/0/3`. The device connected to port `Gi1/0/2` will never see the traffic because the switch knows that the traffic does not go to that device. With broadcast domains, the traffic will still go to all devices connected to the same LAN; this does not change.

Network routing and layer 3

What would happen if you wanted to communicate with someone on a different network or across the internet? You need a device that communicates at layer 3. Layer 3 devices contain routing tables to forward packets from one destination to another. A typical routing table would look like this:

```
C 10.0.0.0 255.255.255.0 is directly connected, Serial0/0/0
C 10.0.1.0 255.255.255.0 is directly connected, Serial0/0/0
O 172.16.0.0 255.255.255.0 [110/128] via 10.255.255.255.1,
Serial0/0/1
```

According to the preceding routing table, there are two directly connected networks and one network being advertised through the **Open Shortest Path First** (**OSPF**) protocol. For instance, if you needed to reach 1.1.1.1 (Cloudflare) or 8.8.8.8 (Google), then it would not know where to send the packets as it does not have those routes. To overcome this, there is the default gateway configuration, which is used to send packets upstream either to a different router in the LAN or to the public internet.

This is great and everything, but what about the layer 3 switches? I am getting there. You see, a typical switch only communicates at layer 2 (we know that), but what if we combined layer 3 functionality with a layer 2 device? This functionality revolutionized the networking field as networks began to grow. This allowed network engineers to purchase switches to route packets to and from various networks with no need for a full routing device.

You will find layer 3 switches in almost every network today because of their functionality. This allows a border router to communicate with the public internet and the LAN, while the layer 3 switches break up broadcast domains into multiple **virtual LANs** (**VLANs**). VLANs allow you to break up broadcast domains but also reduce the number of hosts that can communicate with each other at layer 2:

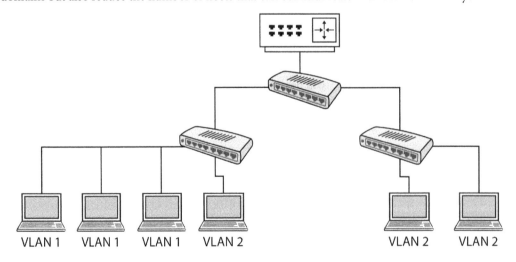

Figure 9.1 – Network design using VLANs

In *Figure 9.1*, we can see an organization's LAN with six computers. There is a router at the top with three switches cascading from it. While this figure is very simplistic, you can see that there are three computers on the left-hand side all connected to VLAN 1, while there are three computers connected to VLAN 2, and that VLAN stretches across multiple switches. With legacy routers and switches, you would see multiple network cables coming from the router down to the switches; now, we only see one. This is due to the 802.1q trunking protocol, which allows VLAN identifications to be carried across switch trunks.

There are plenty of topics regarding networking that we could get into, but this is a very simplistic view of what a network could look like. It is meant to prepare you for upcoming networking topics. Next up, we'll look at IPv4 addressing.

IPv4 addressing and micro-segmentation

Before we get to micro-segmentation, we first need to understand IPv4 addressing. An IP address is much like your street address. It provides a means of routing traffic from one location to another. Every device connected to the network has an IP address, as without it, traffic would never go outside of its local network and the IPv4 address space is in short supply.

Though we ran out of IPv4 addresses years ago, many of us still use IPv4 addressing as it is prevalent throughout the U.S. and much of the world. The utilization of RFC 1918 address spaces helped in dragging out the inevitable: movement toward IPv6.

Traditionally, micro-segmentation is dealt with using multiple VLANs. VLANs, as discussed earlier in this chapter, are a layer 2 concept. This means that IT resources can talk to one another at the data link layer. Networks, especially internal networks, were created to allow up to 254 hosts to freely communicate with each other. That is a lot of chatter. What was OK a few years ago has proven to not be OK today.

Attackers are clever but also lazy – this is true! Once they have gained access to one IT resource on your network, guess what they will do? They will utilize that resource to download their attack tools and get to work. By using that resource, they have now gained access to IT resources in your **demilitarized zone** (**DMZ**). If you have scaled your DMZ network to allow for 254 hosts, the attacker will have a field day.

Micro-segmentation is the theory of reducing a VLAN down to only allowing a certain number of hosts on that network. In *Chapter 6*, we discussed scoping. Scoping is when you evaluate a set of systems that are all together – for instance, PCI-related systems. This is one method of combining systems and placing them in their own VLAN. Another method is through the use of **communities of interest** (**CoI**). CoIs involve taking systems, such as web servers, that are presenting the same content and placing them in their own VLAN (see *Figure 9.2*):

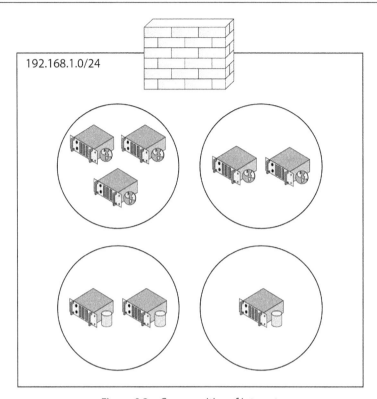

Figure 9.2 – Communities of interest

The interfaces for these VLANs would typically reside on a firewall, or some sort of device that can perform stateful firewalling. The filtering device would then only allow specific source/destination port/protocol traffic from one interface to another. This greatly reduces the blast radius of an attacker gaining access to a system as they would only be able to access significantly fewer hosts than before. Review the following table for micro-segmentation and reduction of hosts:

CIDR	Subnet Mask	Maximum Hosts
/24	255.255.255.0	254
/25	255.255.255.128	126
/26	255.255.255.192	62
/27	255.255.255.224	30
/28	255.255.255.240	14
/29	255.255.255.248	6
/30	255.255.255.252	2

Table 9.2 – Network sizing

The first question you are probably asking yourself is, "Why did it stop at /30?" As /30 can only have two hosts max on a VLAN, /31 and /32 will not allow any hosts at all. By reducing the number of hosts on a VLAN and placing a filtering device at the head end, you have control over what can communicate with the IT resource.

There's no doubt that subnetting is an art. However, it is extremely important when trying to understand how to divide networks. *Table 9.2* depicts micro-segmenting a class C network – but what if you had larger networks? There are a few different tools that you can use to assist with determining the correct size. ipcalc, a free tool used to calculate subnets, is widely available for you to download on macOS and Linux. The following is an example:

```
ipcalc 10.0.0.1/25
Address:    10.0.0.1              00001010.00000000.00000000.0
0000001
Netmask:    255.255.255.128 = 25 11111111.11111111.11111111.1
0000000
Wildcard:   0.0.0.127             00000000.00000000.00000000.0
1111111
Network:    10.0.0.0/25           00001010.00000000.00000000.0
0000000
HostMin:    10.0.0.1              00001010.00000000.00000000.0
0000001
HostMax:    10.0.0.126            00001010.00000000.00000000.0
1111110
Broadcast:  10.0.0.127            00001010.00000000.00000000.0
1111111
Hosts/Net:  126                        Class A, Private Internet
```

As you can see, we have decided to subnet 10.0.0.1/25. ipcalc will then calculate the subnet mask and its wildcard (used by many Cisco-based **access control lists (ACLs)**) and tell you how many hosts you can have on a single network.

As you can imagine, VLAN sprawl became a nightmare for large organizations, and it had its limitations too. A typical network switch can only maintain 4,094 VLANs. This is a problem for larger organizations that have thousands of servers. You can run out of layer 2 space very quickly if it is not planned for properly. It is also a nightmare to maintain and track. A better solution had to be out there.

Virtualization and cloud computing brought about a completely different method of micro-segmentation. While it was difficult in the physical world, things became much easier to control in the virtual world. Many type 1 virtualization platforms – those that can run on bare metal – allowed firewall vendors to create virtualized appliances. As integration between the hypervisor and the virtualized firewall matured, it allowed filtering to happen on the **virtual network interface card** (**vNIC**). This meant that you could maintain larger VLANs – /24 in this case – and still scrutinize the traffic going between systems (east-west traffic). To make things easier, CoIs are grouped together, providing a micro-segmentation-type ruleset.

What about IPv6? The same methodology applies. While IPv6 addressing can go all the way up to ::/128, I often see ISPs and organizations hand out /64 address schemes like they're nothing. It reminds me a lot of the early days of the internet. Organizations, higher education institutions, and even my grandma could have applied for and received a network capable of having more than 16 million hosts in the 1980s and 1990s. Organizations such as Apple and Ford Motor Company have these extremely large networks. However, as we quickly came to realize, the number of gadgets needing an IP address outnumbered the availability. Will the same apply to IPv6? Yes, yes it will.

Next, we'll take a look at network security. As you will see, traditional network security was very simplistic or did not exist at all. As we continue to improve network security, we look at micro-segmentation being used in networks that heavily relied on IPv4 addressing schemes, as previously mentioned.

Micro-segmentation does not have to only be applied through subnetting and firewalls; many cloud service providers offer this layer of security as well. For example, **Amazon Web Services** (**AWS**) creates security groups inside a **virtual private cloud** (**VPC**). Each server, or EC2 instance, can be placed inside its own security group, protecting itself from other systems inside the VPC. As this is performed inside the VPC, it provides additional security layers to protect your IT resources. Like micro-segmentation, you can have as many systems as you like inside a security group, though it is recommended to minimize that amount for better protection.

Traditional network security

Network security has evolved over the decades. Traditionally, a network had a single firewall at the perimeter (or multiple for high availability) that was used to protect an internal network. This scenario was also the case if the organization had servers that were externally exposed to the internet through a DMZ setup. This type of configuration was known as *hard outer shell with a softer middle*. Today, that border is disappearing. Cloud computing and a remote workforce have eroded any resemblance of a border network for an organization.

During the 1990s and 2000s, many organizations had not deployed a firewall for their organization, simply relying on their border router to route traffic without filtering. Some organizations that had deployed a firewall in their network had *allow all* rules, preventing nothing and acting as a *bump in the wire*. As adversaries became more sophisticated, more organizations deployed a firewall, but the rules became *allow by default, deny by exception*. Organizations knew they needed to have a firewall in place, just no clue how to correctly configure it.

Throughout the 2000s and into the 2010s, regulatory bodies began to mandate that organizations have a firewall in place. As things progressed, and these regulatory standards became more defined, the rules began to change. Regulators now required specific types of firewalls to be used in the environment with *deny by default, allow by exception* rules. While this enhanced the security posture of the network, many in the IT field saw this as a burden.

Applications were written without much security in mind, if at all. This also meant that we did not fully understand how the application communicated with others either in the network or across the internet. This posed a large problem in that network and firewall administrators could not protect the application adequately enough. We continue to run into this same problem today, though it is far less frequent. This was also a time in the history of IT and security when many relied on the network for securing their applications and the rest of the organization.

Information security practitioners had to truly layer on the security products to fully understand what was going on in the network. First, there were firewalls, then came IDSs/IPSs, DLP, and load balancers. These could have all come from the same provider, but many organizations purchased the best that they could afford at the time. As CPUs and memory became cheaper, and transistors became smaller, manufacturers began to lump all that security goodness into a single appliance. This began the movement toward **unified threat management (UTM)** firewalls.

UTM firewalls were a different breed altogether. First, they combined all the technological security. Second, they allowed security administrators to have a single-pane-of-glass view of their environment. Third, there was less to maintain, which lowered capital expenditure. The downside was that companies had to purchase larger appliances than they previously had in their environment.

Firewalls were traditionally sized based on the throughput of the internet speed. While the same is true today, you must consider the additional features that you plan to activate. Features take up CPU and memory, as do **deep packet inspection (DPI)** and **network address translation (NAT)**. To use these add-ons, you must buy beefier boxes than you would normally buy to perform these types of functionalities. As you will see in the next section, regulators do not always see eye to eye when it comes to security.

Traffic inspection

During the mid-2010s, we also began to see a real drive toward data privacy and security. Regulatory bodies were mandating that sensitive network traffic be encrypted across the public internet. As the years went on, it became more common for internal network traffic to also be encrypted. Adversaries were also becoming more sophisticated and encrypting their payloads and **command and control (C2)** traffic, blinding IDS/IPS and UTM firewalls. Search engine optimization and web rankings in online searches now mandate that your public-facing website be encrypted or lose significant rankings.

A former company I worked for was undergoing its annual PCI audit. When the auditors came onsite to perform their inspection, they took a hard look at the security stack. During their inspection, the auditors uncovered that the company had been performing SSL offloading on a device that was used to pass traffic. This drew a red flag from the auditors, which caused them to find something. They mandated that no device, regardless of its capabilities, could decrypt SSL traffic between the user and the server.

Mind you, this was a special-purpose security device. To properly maintain network flows, the device had to decrypt it. The manufacturer of the network device was security focused in that special internal hardware chips were created to perform this type of inspection with encrypt/decrypt functions happening in milliseconds. All that security did not matter to the auditors. In my opinion, having the device not perform SSL offloading to inspect the traffic lessened the security posture of the environment. There is a point to all this in that to perform DPI, you must decrypt network flows, and this is not easy.

Network taps, SSL/TLS offloading devices, and server agents can all perform the decryption functions of network flows. Without this, even the most expensive security equipment you can buy will not see your network traffic. As previously mentioned, adversaries are also encrypting their network traffic to bypass security mechanisms. When a client attempts to make an encrypted connection to the remote server, that server holds onto the private key. At no point in time does (or should) anything else know how to decrypt that traffic. All you and your network equipment will see is `TCP:443` (typically) going to and from the network.

Decryption has almost become mandatory from a network perspective. As we will discuss later in this chapter, firewalls are now capable of detecting and blocking malware in the network stream. They are capable of detecting **Domain Name System** (**DNS**) traffic and blocking the web connection. Firewalls can now understand the difference between a personal and corporate Gmail account. This would not be possible without decryption.

I get privacy. I know privacy. I have been a privacy engineer and a data privacy officer. There are additional measures that you can put in place to maintain the privacy of your users while still performing network security inspections. Website categorizations are one way of maintaining privacy. Manufacturers have built-in feature sets to allow you to decrypt traffic while maintaining encryption for HIPAA-related sites for example. Create and maintain privacy notices and policies for your employees and stakeholders. Adopt privacy frameworks such as the **Generally Accepted Privacy Principals** (**GAPP**) and have regular assessments to show proof. Develop a **privacy by design** (**PbD**) methodology and ensure that it is being performed for every project.

Networking is a growing field and is quickly becoming disrupted by how traditional networks are being built. Networks today rely on automation and **infrastructure as code** (**IaC**), which are all built upon code.

Networks of today

Networking today is evolving rapidly. What was once decentralized administration of network equipment has moved toward single-pane-of-glass administration. Software-defined networks have grown into the use of scripting languages to automate network creation and administration. We are moving toward IaC and, more importantly, compliance as code. Companies are hiring software engineers to perform network functions and have created an entirely new field of **site reliability engineers (SREs)**.

The network field is undergoing a major transformation. Traditional network administration consisted of secure shelling or telnetting (I hope not anymore) into a network switch or router and plugging away at typing commands into a terminal window. Once the network administrator had completed their task, they saved the configuration and logged out. The larger the organization, the more network equipment had to be configured by hand. This meant it was difficult to standardize network configurations across hundreds of network equipment throughout a campus.

In the early 2010s, **software-defined networking (SDN)** began to take off. This allowed network administrators and software engineers to write code against their networking equipment. Open source projects such as OpenDaylight pioneered SDN development to make network automation what it is today. Talk of SDN has all but faded away, but there are new ways to automate network configurations.

What if you were to learn one programming language that could allow you to configure your network equipment, regardless of the manufacturer? How about taking that language and also applying it to the cloud? What about creating new servers, all from the same code base? Why, that almost sounds too good to be true! But wait, there's more… Not really, but if my infomercial did not entice you, what if I told you that you could do all of this for free?

IaC has been a driving force for network automation over the last few years, especially with cloud computing. Applications such as Red Hat's Ansible and HashiCorp's Terraform have become the predominant forces in automating network configurations. With IaC, all you need is a little bit of networking and scripting knowledge, and you can begin creating cloud-based networks or work with on-premises networking equipment.

Regulators also want to ensure that every network change is tied back to an approval process. In most instances, this comes down to having a work order that depicts the changes that were made against the network, especially when it comes to firewall changes. By using IaC and tying that back to a **version control system (VCS)**, you maintain historical changes throughout the life cycle of that environment. For instance, take a look at the following code block:

```
resource "aws_vpc" "example_corp" {
    cidr_block = "10.17.0.0/16"

    tags = {
        Name = "Corporate VPC"
        WorkOrder = "109650"
```

```
        }
    }

    resource "aws_subnet" "websrv_subnet" {
        vpc_id = aws_vpc.example_corp.id
        cidr_block = "10.17.17.0/28"
        availability_zone = "us-east-1a"

        tags = {
            Name = "Main corporate web server subnet"
            WorkOrder = "109651"
        }
    }
```

In these 19 lines of code, we have created a VPC in AWS, along with a subnet dedicated to web servers. We set aside a fairly large subnet mask in which smaller networks are created. Then, it associates work orders 109650 and 109651 to each stanza, respectively. Once you are done with the code, next, you must commit this to a VCS such as Git. This will help you not only track the changes being made but also maintain compliance by associating already-approved work order numbers with each group. This will make it easy to understand who wrote the code, who applied the changes, and who approved the changes.

The networking field has changed over just a few decades, from specialty equipment with manufacturer-specific commands to configure the equipment to having a common code set that can be applied to multiple manufacturing types at once. We often talk about digital disruptions in various markets, with the largest disrupter being the internet in general.

Now that we have our network built, we must begin to build trust. Building trust is not building confidence in the network; it is whether we trust the devices connecting to it.

Building trust into the network

Traditional networks operated by implicitly allowing anyone to connect to the network. IT had complete control over every device in the corporate network, which meant that they could trust these devices. As laptops became cheaper, and we became more mobile, trust in these devices dwindled. Years ago, using a device anywhere was (mostly) acceptable as mobile computing had not matured enough and was not cheap enough. However, within the last few years, **bring your own device** (**BYOD**) and working from home have eroded trust in devices.

How do we begin to trust all the various devices that connect to our network? How do we know whether the person who is logging in to their corporate account is who they say they are? Can we know whether the device connecting to our SaaS mail provider is corporate-owned, a personal machine, or an airport kiosk? This all comes back to the level of risk your organization is willing to accept.

Traditionally, building trust in the network was hectic. If you worked at large organizations or higher education institutions, you were accustomed to splitting the network. Large organizations had a DMZ not only for their public-facing applications but also for untrusted devices. Untrusted devices are described as those devices over which you have no control. This includes BYOD and vendors' or salespeople's mobile devices. Higher education, on the other hand, is a unique environment to work in as you wear two different hats. First, you must protect the business. Without the business, you have no institution. Second, you act as an ISP. As many students live on campus, internet connectivity must be accessible. Both require 24x7 support and different policies on how you run the network. I would dare say that higher education felt the initial brunt of BYOD and was mostly unprepared for it.

Having hundreds, if not thousands, of BYOD devices all connecting to your network LAN and **wireless LAN** (**WLAN**) at the same time and bumping up against business-related devices spelled disaster for many. In the early 2000s, we saw Code Red, SQL Slammer, and Blaster bringing many networks to their knees. This was on top of students having access to unlimited bandwidth and fast internet speeds, which brought about Napster and other peer-to-peer file-sharing services.

If the student side was bringing down the network, how could the business side do their work? The answer was that they could not be productive. This led to the inception of **network access control** (**NAC**). NAC brought about many network positives in that you could then force untrusted devices off the network until they had accepted the organization's network policies. This traditionally meant that your device had to conform to the following:

- An agent installed on the computer

- An updated operating system

- Installed and maintained antivirus software

- Removed P2P software

If you were a student on campus and did not have these configured on your machine, that meant no network for you. While we had to protect ourselves from the kids running with scissors, this stunted the computer science industry. This forced many institutions to enforce draconian policies on their students. Some would only allow Windows and some macOS computers, which meant students could not use UNIX or Linux based systems. Some higher education institutions made exceptions to this, while many did not.

Until you had the updated security software installed and configured, you were placed into a dead-end VLAN. This was a VLAN that either did not have a network interface or had very restricted ACLs placed on that network. Once the agent validated that the machine had the necessary security configurations in place, it allowed the machine to hop over to a network that had access to the internet.

While this was a monumental step in securing the network, it was not without its flaws. Today, agent-based software will restrict network access or isolate the device from the network. Traditionally, however, these agents relied heavily on network configurations. Many NAC products either could not work with certain devices or they would work but it was painfully slow.

Many solutions relied on sending and receiving SNMP read/write commands from across the network. As you could probably imagine, this introduced a large security flaw in the network. Solutions back then did not rely on SNMP version 3, which allowed for authentication and encryption. SNMP version 2c allowed for additional functionality over version 1, but it still relied on a single community string to be sent across the network, unencrypted. This meant that students could have also gained access to those community strings. What could possibly go wrong?

While NAC has not gone away, it comes in many different forms, one of those being **mobile device management** (**MDM**). As BYOD began to take hold of corporate environments, they too needed a way to trust devices connecting to their network, and MDM was that solution. It allowed corporations to place policies on mobile phones, tablets, laptops, and desktops. As mobility is the driver for an MDM solution, it became SaaS-based, whereas NAC is typically an on-premises solution. MDM policies include having the following:

- An agent installed on the device
- Updated antivirus
- An updated operating system
- Hard drive encryption
- Screen lock with a defined passcode length
- Protections against it being jail-broken

These configurations had to be present on the device before it could access the internal network and, in most cases, many cloud-based applications as well. MDM also began the process of making the network recognize the mobile device. Network administrators could build profiles for enrolled devices to ensure they met policy. There was also a security feature that was not present in traditional NAC solutions – it allowed for remote wiping of the device. If an employee were to ever lose or have their device stolen, an administrator could send a signal to the device, telling it to wipe the hard drive.

This also meant that if the device did not have hard drive encryption or other configuration items set per the policy, the device would be restricted from accessing sensitive material. This required layering extra security onto the device. As you continue to layer on security configurations, you lower the overall risk to the data that resides on it. Many think that information security is about how many bells and whistles we can activate, but it isn't. It is about securing the data that resides on an IT resource.

Many configuration policies were automatically applied to the device upon installation of the agent and acceptance of the policy. As this automatically configured many of the configuration items, it reduced the number of calls coming to the help desk for assistance. MDM was traditionally thought of as a BYOD solution, meaning it was not meant for on-premises corporate devices.

Microsoft is making a bet on the fact that this is the future of endpoint management and security. With the introduction of its own MDM solution, Intune, Microsoft is making the push for an all-in-one solution. This solution not only includes mobile phones and tablets but also is now also the preferred management solution for Windows 11.

Building trust must account for those connecting to your LAN and from a remote location as well. As the world is becoming a more *mobile* society, today's workforce does not necessarily go to the office regularly. We must account for those who are across the country, or the world, who work for the organization.

Virtual private networking and remote access

As we became mobile, employees needed a way to securely connect back to the internal corporate network. Organizations that had more than one facility also wanted a secure way of sharing IT resources. In some cases, dedicated circuits between facilities were too expensive and required expertise that many organizations just did not have. **Virtual private networking** (**VPN**) grew out of these needs.

The first VPN protocols were heavy, reducing the amount of throughput you could achieve over an internet connection, but they got the job done. One of the oldest protocols, IPSEC, runs at layer 3 of the OSI stack. Though the oldest and heaviest, it is the most reliable type of VPN connection used today. Even with the onset of newer VPN protocols, such as SSL/TLS VPN, it is often used over any other as IPSEC runs at layer 3 and secures data the rest of the way up the stack. See, I told you the OSI model would come in handy!

As SSL/TLS became prevalent in terms of securing network connections, it too became a heavily favored protocol. Though it had many issues early on, SSL/TLS VPN became the de facto standard for many remote VPN connections. Issues with SSL/TLS VPNs were prevalent for many different protocols such as database traffic. This was due to the fact that SSL/TLS VPN networks connect at layer 4, just one layer up from IPSEC.

Companies also began to perform posture checks on remotely connected devices. This was rarely done, however, because it added complexity to the VPN connection itself. Though limited in capabilities, these posture checks ensured that the device connecting had updated operating system software, antivirus, and more. If posture checks were being performed, they were only being performed once, upon initial connection to the VPN device.

Once the connection was established, the edge device had two different ways of routing traffic across the internet: full or split tunneling. Full tunneling treats the traffic the same – it routes all traffic across the internet, to the corporate network, and out the company's front door. In split tunneling, all traffic destined for the corporate network is routed through the tunnel, whereas anything else goes out the user's ISP. Full tunneling poses a problem in that the ingress/egress traffic of the corporate network is being consumed for every remote VPN connection. If you have 500 concurrent VPN connections, all that traffic is routed through the corporate network. Split tunneling poses a problem in that it does not fully protect the edge device.

Anyone who has ever gone to a security conference knows that you have to have your electronic shields up. While many attend to gain knowledge from others, there are a few that like to play around; you know, for educational purposes only! That means your level of trust in the network you are about to connect to becomes little to none. You want that full tunnel connection to the internet to ensure that your traffic is not being sniffed.

Many medium- to large-sized organizations segmented their networks in such a way that firewalls were placed between VPN concentrators and the rest of the network. These VPN networks were split based on the level of risk of the edge device connecting to the network. For instance, there was the creation of an extranet, which further reduced what a third party could connect to, whereas an employee concentrator had almost full access to the internal network.

Building trust in the network is a great way of securing the network. When you begin to build trust, you start to understand who is connecting to the network. Much like a firewall should start by denying all traffic and only allow by exception, so too should we take that mindset of those connecting to our environment. This is why the U.S. federal government, and many public/private organizations, are beginning to utilize zero trust.

Getting to know the "zero trust" concept

What is zero trust? It is a concept that says that unless you can verify the security configurations and the user of that device, it is considered untrusted. Zero trust also means that we are not going to just verify that it is the correct user on the initial login. We are going to continuously validate the user and their device throughout its connection to corporate IT resources. We are also going to layer on security configurations, much like what NAC and MDM provided, and ensure those configuration items are present on the device.

First, we identify the identity being used to connect to the corporate network. This comes in many different flavors, but the most common way is by connecting your zero-trust agent to your IdP. In addition to forcing our employees to use their corporate username and password, we also force the user to use multifactor authentication, which further identifies the user. Once it has been validated that the user is who they say they are, we move on to the next step, which is validating the edge device.

Much like MDM, we build policies to enforce the security configurations of the edge device. This can be configured by using the same policies we used previously; however, we can do much more. This includes installing corporate-owned certificates on edge devices to ensure these devices are not personally owned and performing posture checks, much like we did with NAC. However, these are automated upon acceptance of the policy. We can also perform policy-based routing, forcing traffic destined for a corporate-owned G Suite or Microsoft 365 instance through the organization's network, while a personal Gmail account goes out over the public internet. In addition, we can create policies that restrict what the user can access, such as web filtering and monitoring DNS traffic.

We can also monitor and block malicious traffic and payloads through a zero-trust agent. As web traffic is proxied before it's received on the edge device, it is checked for malicious activity. This means that the traffic must be decrypted for the proxy to see it. While there are plenty of privacy concerns, it will require the edge device to install a certificate that is mutually agreed upon to perform that function. Through the zero-trust agent, we validate and re-validate at given intervals. This means that if the device fails a policy at any point, it becomes untrusted.

Zero-trust agents rely on the identity of the user connecting to the network. Tied back to role-based access, employees are segmented from vendors and contractors. Policies are then created, allowing or forbidding that user to access internal and external resources.

Firewalls are one of the ways that an organization can easily implement the zero-trust concept in its environment. Hence, it's important to know the functionality and the types of firewalls you can use to your advantage.

Understanding firewall functionality

Have you ever heard the phrase "If it's not the network, it's the firewall"? It seems like every time I turn around someone is blaming the firewall for their issues. The application doesn't work? It's the firewall. The internet's moving at a crawl? It's the firewall. Kids won't listen to you? Blame the firewall!

There seem to be firewalls for just about everything – firewalls for network traffic, firewalls for applications… there are even specific firewalls for DNS traffic. There are plenty of best practices available to protect your internal traffic from the internet. Firewalls are also best known for NAT and **port address translation** (**PAT**) (yes, there is a difference). There are also differences between host-based firewalls and all the rest:

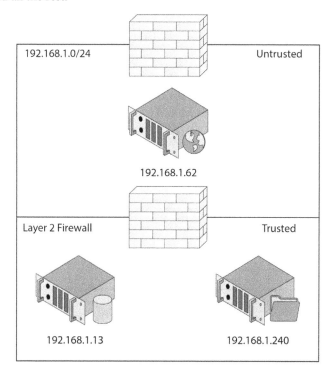

Figure 9.3 – Layer 2 firewall design

There are plenty of different types of firewalls out there, and they work at just about every layer of the OSI model. While typical firewalls operate at layer 3, the network layer, you will find layer 2 firewalls stashed away in many places of organizational network architecture. Certain applications may require maintaining the network interface on a separate device other than on a firewall or router. When placing a firewall into a layer 2 mode, you can gain some functionality, but lose a lot.

In certain LAN and WAN configurations, a layer 2 firewall is a must. Much like the traditional layer 3 firewall, it uses rules-based ACLs to filter traffic to and from the IT resource. The firewall also maintains a trusted and untrusted side of the device. The untrusted side of the device resides on the same VLAN as the network interface, which allows traffic to route out of the network. The firewall, however, performs a VLAN translation from the untrusted side to the trusted side. That translated VLAN ID is used on the trusted side to protect all IT resources. For instance, the firewall knows of two different VLANs, 17 and 117. VLAN 17 is the untrusted VLAN, whereas 117 is on the trusted side. All devices maintain the same layer 2 functionality (though split between trusted and untrusted networks) and still share the same network interface.

IT resources on VLAN 17 do not have any rules-based filtering applied to them, essentially meaning the ports are wide open to the rest of the network or the internet. However, IT resources that reside on VLAN 117 are protected by the firewall. This means that these resources are protected by filter-based rulesets. The win here (and it is the only win) is that all IT resources are utilizing the same layer 3 interface.

When placing a firewall in layer 2 mode, you lose most of the firewall's functionality. The firewall is no longer capable of creating VPN connections, both site-to-site and client-based. You also lose the functionality of having the firewall perform NAT-/PAT-based translations. It just becomes a layer 2 filtering device – that is it.

In layer 3 firewalls, you gain all of that functionality but you are not able to filter traffic based on layer 2. This means that the firewall has now become the network interface for the network behind it. It also means that you can create additional layer 3 interfaces, which are sometimes needed for remote VPN connections. Lastly, you can perform NAT-/PAT-based translations.

Many U.S. based networks rely on the RFC 1918 address space. This address space is not meant to be routed across the internet's backbone. There are three different types of RFC 1918 address networks. They are as follows:

Class of Network	Network Size	Number of Hosts
A	10.0.0.0/8	16,777,214
B	172.16.0.0/12	65,534
C	192.168.0.0/16	254

Table 9.3 – Network class

Many home networks utilize the `192.168.0.0/16` address space, whereas extremely large networks use the `10.0.0.0/8` network size. This is also a big reason, as stated previously, why the U.S. and many other countries have yet to fully adopt IPv6. However, how can we route 254 addresses out of a single publicly routable address or multiple addresses? To do that, would the organization also need 254 public IPs? Not so fast!

NAT allows us to map an internal IP address to an external IP address. This allows us to control which IT resources can route out to the internet. Static 1:1 NATs are also common for accessing internal resources from an external IP. This allows us to, again, map an external IP to an internal IP, and then pass that traffic to our internal network. This would effectively hide our internal address space. As you can imagine, though, this can be quite expensive as there has to be a 1:1 mapping between the two.

This is where PAT comes in handy. You likely use or are familiar with PAT; you just may not know about it. For instance, your internet connection at home likely uses PAT as you only have one external IP address, which resides on a cable modem or router. With NAT, you could only have one internal device on your home network to route out at any given time, but what if you have 5, 10, or 1,000 hosts that all need to route out?

When you make an outbound TCP connection from your computer, it will create a source port for it to use. That source port is then used to create a socket connection to a remote resource across the internet. There is also a default TCP port that is used on the receiving end. For TLS web traffic, the default is `443`. A router or firewall also maintains a state table that keeps track of each outbound connection. That state table maintains a `key:value` pair for the source and destination. This can look like this:

```
10.0.0.65:49953 TCP à 1.1.1.1:53 TCP
10.0.0.65:50014 TCP à 23.203.117.159:443 TCP
10.0.0.72:49950 TCP à 142.251.32.4:443 TCP
10.10.1.156:64532 TCP à 157.240.245.35:443 TCP
```

In the preceding example, it appears that we either have two smaller networks or one very large network. This does not matter as the routing device will be able to route all that traffic outbound to the internet. You could have one to however many networks you want behind the firewall; if the rules are set up properly, it will route out. On the destination side, however, it looks a little different.

As you have probably guessed, if the traffic is being routed out through a publicly available interface, would the traffic not have to come back to that same interface? Yes, or we would have created bad network routes. If our public IP were `8.8.8.8` and was used for PAT, then all the destination would see is `8.8.8.8` and not our internal `10.0.0.65` address.

Much like the rest of the industry, firewalls are also going through a transformation. Traditional firewalls only looked at the source/destination/port type of traffic. They never cared about the type of application that was set behind it. For instance, most networks allow ports `80` and `443` to be outbound. If they didn't, then all web traffic would stop. So, we have `80/443` outbound but restrict the **Secure Shell** (**SSH**) protocol, because we do not need remote administration on the internet. If we were to change our SSH daemon from listening on port 22 and moved it to 443, then that would work – defeating our network policy.

To combat this, firewall manufacturers began to develop application-aware firewalls, also known as layer 7 firewalls. Protocols typically have banners or a way for the application to know and interact with its destination. A web server banner can look like this:

```
HTTP/1.1 200 OK
Content-Type: text/html; charset=ISO-8859-1
P3P: CP="This is not a P3P policy! See g.co/p3phelp for more
info."
Date: Sun, 04 Sep 2022 18:03:06 GMT
Server: gws
X-XSS-Protection: 0
X-Frame-Options: SAMEORIGIN
Transfer-Encoding: chunked
Expires: Sun, 04 Sep 2022 18:03:06 GMT
Cache-Control: private
```

On the other hand, an SSH banner can look like this:

```
SSH-2.0-OpenSSH_8.7
```

As you can see, the banners are quite different. The web server banner has much more output than the SSH banner. An application layer firewall will be able to see that traffic and filter it. If an SSH daemon were listening on port 443:TCP and we tried to make a connection, the firewall would block it as the firewall is expecting to see a web-based banner, not an SSH banner.

Web application firewalls

Web application firewalls (**WAFs**), while still operating at layer 7 of the OSI model, protect inbound connections directed toward web applications. They perform functions that mitigate many web-based vulnerabilities today. The **Open Web Application Security Project** (**OWASP**) has created the OWASP Top 10. It specifies the top 10 most prevalent vulnerabilities discovered in web apps today. Its latest version, 2021, has the following listed as its top risks:

ID	Misconfiguration
A01	Broken Access Control
A02	Cryptographic Failures
A03	Injection
A04	Insecure Design
A05	Security Misconfiguration
A06	Vulnerable and Outdated Components
A07	Identification and Authentication Failures
A08	Software and Data Integrity Failures
A09	Security Logging and Monitoring Failures
A10	**Server-Side Request Forgery (SSRF)**

Table 9.4 - OWASP Top 10

The full list can be found at `https://owasp.org/www-project-top-ten/`.

In addition to blocking attacks that the OWASP Top 10 describes, many managed WAF providers also have mitigations for popular web frameworks and **content management systems** (**CMSs**). There are pre-built rules that can be used to mitigate vulnerabilities in development frameworks such as ColdFusion, and the most popular CMS on the web today, WordPress.

You do not have to spend money on a WAF – you can attempt to build one yourself using Apache's ModSecurity. ModSecurity is a rules-based engine that allows application and network administrators to create rules to block web traffic. You can leverage pre-built rules from OWASP[1] and/or through Trustwave's GitHub repository[2]. I highly recommend that if you plan to roll out a WAF, you take the time to understand your web application and how it will interact with the various available rulesets. Otherwise, it could accidentally make your web application unusable or not function at all.

DNS firewalls

Another add-on for security came with the introduction of DNS firewalls. There are plenty of known malicious websites on the internet, but how can you keep track of them all? One way of doing this is by purchasing a managed service that will keep track of this for you, and again, there are plenty of free services available to you.

As we now know, firewalls are great at many different things. However, they don't know what is good or bad as their number one job is to filter traffic. We have to layer yet another piece onto our network, blocking access to these sites. This is where DNS firewalls come in.

DNS firewalls accept and process recursive lookups for internet-based traffic. To perform these lookups, you must have your DNS settings set up to forward all DNS traffic to this recursive lookup service. Once configured, the managed service provider performs all DNS lookups on your behalf. If your local machine were to perform a lookup against a known malicious website, the managed service provider would spoof the correct IP address with one that it knows is trusted. Once your machine has that information and goes to make a connection, you will be presented with a page that tells you that the traffic was blocked because of a policy issue.

As an example, you receive a phishing email in your inbox. This phishing email states that your PayPal account has been hacked and you need to immediately change your password (sound familiar?). We figure that the link is legitimate as it came from `account-help@paypal-accounts.com`. I mean, why would we not trust it? It has PayPal in the domain name, and they are PCI certified!

Anyway, the service knows that the domain is malicious and instead of presenting you with the phishing website, the service presents a page that says **This website has been blocked by your corporate policy**. This not only protects the user from doing something they should not do but also protects the organization. These attempts against websites can be tracked and monitored as well, alerting the organization's **security operations center** (**SOC**) to take action in the event this occurs.

Many firewalls today are capable of filtering traffic based on the application type and certain traffic signatures. As the head of security, you should understand what the business needs to purchase and implement the technology appropriately. Service providers such as Cloudflare, Quad9, and others provide DNS-based firewalls. It may make sense to review cloud-based service providers to see what they have to offer rather than building or implementing your own.

DNS attacks are very common due to insecure configurations. These insecure configurations allow for recursive lookups from anywhere across the internet. One such attack, DNS amplification, is when the attacker spoofs the source IP, placing the target's IP address as the source. Do this across a few thousand machines and you end up with a **distributed denial of service** (**DDoS**) attack.

Distributed denial of service

DDoS is a method of resource exhaustion. A DDoS attack is meant to utilize all possible resources against a system, which can include bandwidth or TCP connections. For example, GitHub was hit by a 1.35 **terabits per second** (**Tbps**) DDoS attack in 2018. This type of attack was meant to saturate the network connectivity of the popular Git SaaS hosting company.

While the GitHub attack was meant to fill up the network, Cloudflare thwarted two separate DDoS attacks against its customers. In 2021, the popular internet-based security company mitigated a 17.2 million **requests per second** (**RPS**) attack[3]. Not to be outdone, the following year, the company was able to stop an attack producing 26 million RPS[4]. This attack originated from a botnet totaling more than 5,000 internet-connected devices. The attack was meant to soak up resources on servers providing content to their customers.

While not necessarily a flaw in the way TCP and UDP work, various DDoS attacks do take advantage of the way they operate. TCP provides a three-way handshake when negotiating its communication. These attacks take advantage of that handshake by only sending the initial SYN packet to the IT resource. These SYN packets are meant to open up communication between the client and the server. If you send too many SYN packets, the server cannot take any more requests and stops serving new customers.

UDP is a *send-it-and-forget-it* protocol. Where TCP waits for the customer's device to acknowledge and respond to the server, UDP does not care. There are many different applications for the UDP protocol, such as the **Network Time Protocol** (**NTP**) and the popular DNS, as it is lightweight and we do not need to acknowledge its receipt. However, botnets utilize this protocol for DDoS attacks as you can send hundreds of thousands, if not millions, of packets to exhaust a resource or take up all the network bandwidth of the victim.

There are a few different ways to mitigate such an attack. While there are free services available, many require you to pay for the protection. Though limited in functionality, Cloudflare provides a free tier that can be used to protect your internet-facing applications from various types of attacks. Some appliances can be installed on the corporate network that can be used to limit the number of packets coming into the network; however, by the time it hits your network, it could be too late.

SaaS-based DDoS mitigation solutions are meant to block all malicious traffic before it reaches its destination. How do they do this? By routing your network traffic through the SaaS provider's *scrubbing* centers. These centers are in different locations throughout the world and clean up any traffic before routing it to the destination. This means that any attempt to attack a client will go through the SaaS provider's network first. Any attacks, either network- or application-based, are stopped before they reach your destination.

Summary

The internet is a fun place to communicate with friends, share photos and videos, and conduct business. It has evolved from two machines trying to communicate with each other to 30 billion connected devices by the end of 2022. It allows us to bank, utilize cryptocurrency, and perform stock trades. It can also be a scary place if you are not protecting yourself properly.

As you have learned in this chapter, traditional networking had us logging in to individual network equipment for configuration. This made network administration a long, drawn-out process when performing equipment refreshes or installing equipment in a new building. Today, we can configure large network stacks with Terraform and Ansible.

There are differing methodologies we can follow to protect us from cyberattacks. Micro-segmentation and zero trust are just two ways to protect ourselves. Reducing the blast radius in the number of servers that can be attacked and continuously revalidating those connecting to our network are just a few different ways to ensure we trust the devices connecting to our network.

The OSI model, as discussed, can help us determine the types of safeguards we can put into the network. Layer 2 firewalls provide protections where an interface cannot be migrated to a router. Layer 3 firewalls have additional functionality, such as providing NAT and VPN, in addition to protecting the network. Layer 7 firewalls provide the most protection as these look at how the application interacts and responds to the user.

Additionally, with the number of different firewall vendors on the market, with varying types of firewalls themselves, how do you choose the correct one for your organization? Reiterating what you have learned in this chapter, first understand how the business generates revenue and what the riskier assets are. If they are publicly exposed, the level of risk to that asset is higher than one that is sitting behind two firewalls and a NATed interface.

Finally, we discussed DDoS attacks that take advantage of IT systems and network bandwidth by consuming all their resources. To prevent this, look into prevention services that will route traffic through scrubbing centers to clean up malicious traffic before sending it to your network.

In the next chapter, we will focus on server-based security. We will look at Windows- and Linux-based operating systems and what you can do to protect them.

References

1. The OWASP mod_security project: `https://owasp.org/www-project-modsecurity-core-rule-set/`

2. mod_security configurations: `https://github.com/SpiderLabs/ModSecurity`

3. Yoachimik, Omer. *Cloudflare thwarts 17.2M rps DDoS attack — the largest ever reported*: `https://blog.cloudflare.com/cloudflare-thwarts-17-2m-rps-ddos-attack-the-largest-ever-reported/`

4. Yoachimik, Omer. *Cloudflare mitigates 26 million request per second DDoS attack*: `https://blog.cloudflare.com/26m-rps-ddos/`

10
Computer and Server Security

CAD drawings, customer records, transaction history, and so on are examples of information that an organization holds onto. This information is important to your business and must be safeguarded from adversaries and your competition.

Computers and the software that runs on them are vulnerable – there is no way of getting around that. However, you, as the head of security, must know what to do to protect these systems to the best of your ability.

There is no way of getting to zero risk – it is not attainable. However, IT resources can be configured in different ways to secure them. They can be patched against vulnerabilities. Encryption can be used to further secure remote access or protect the information that resides on a system.

The business will want you to secure these devices to the best of your ability. Does it seem like a big task? It can be. The majority of software providers make it easy to patch a discovered weakness. Microsoft has Patch Tuesday, while many Linux distributions provide updates that can be applied immediately after release. This chapter will review what to do to ensure your operating system environment is secured and maintained – applying software patches to software and hardware (yes, hardware is important too), remote management, server hardening concepts, authentication, and understanding the **Internet of Things (IoT)**.

In a nutshell, the following topics will be covered in this chapter:

- The history of operating systems
- Exploring server hardening steps
- Dangers of TOFU
- The IoT
- Understanding encryption

The history of operating systems

While the internet has been the cause of many market disruptions in modern history, we would not be anywhere without computers. Electronic computing systems have been around since the 1940s. During that time, computers were massive in size, often taking up large warehouses. Computer programs were written in machine language and fed to the computer using punch cards.

The first operating system, **General Motors Operating System** (**GMOS**), was released in the 1950s. GMOS was created for the IBM 704 computer. Specialty operating systems were created throughout the 1950s and 1960s. In 1971, the first UNIX operating system was created by Dennis Ritchie and Ken Thompson at the Bell Labs research facility. Around the same time, the programming languages C and Assembly were developed.

One of the most influential operating systems, Microsoft, was founded in Albuquerque, NM in 1975 by Bill Gates and Paul Allen. A year later, Apple co-founder and longtime CEO Steve Jobs and Steve Wozniak created the Apple-1. This was the first example of modern-day computing for home use. The sale price for an Apple-1 at the time was $666.66. That would equate to $3,175 if it were sold in 2021. Although less expensive than the massive computers used by corporations, it was still too expensive for general home use.

As computers matured over the years, their sizes began to shrink and they had more processing power. During the 1980s, mobile computing started to take off with the first laptops becoming available on the market. Although many began to see computing as more than just business-related machinery, computers were still too expensive for many. It was not until the early to mid-1990s that we saw computers in the home.

Software and operating systems were looking for a change as well. The **disk operating system** (**DOS**) was released in 1981 by IBM and Microsoft. Microsoft then released multiple versions throughout the 1980s. 1984 introduced the first Macintosh operating system. This revolutionized the home computing industry, as it introduced a **graphical user interface** (**GUI**). Instead of being command-line-driven, the GUI presented a user-friendly interface for non-hobbyists. A year later, Microsoft released its first version of the Windows operating system.

At the time, much like how computer hardware was expensive, so too were operating systems. Frustrated by how expensive computer software was, a computer science student from the university of Helsinki, Finland, created his own operating system based on Minix. Minix was a UNIX-based operating system used for software development at the time. That computer science student was no other than Linus Torvalds, the developer of the Linux kernel.

While Apple was focused primarily on home use and Linux was just starting to take hold of the market, Microsoft had developed its first real operating system meant for business – Windows NT 3.1. Released in 1993, Windows NT 3.1 was for sale just a year after Windows 3.1 was released for home use. Arguably the most important Microsoft Windows release, Windows 95, revolutionized the computer industry. It was the first real operating system that made accessing the internet easy. This also introduced Internet Explorer, one of the first graphical internet web browsers.

Throughout much of the 1990s and 2000s, wars seemed to rage on with the debate between open source and closed source operating systems. Much of that debate centered around the security of the software. Proponents of open source software believe that the more people you have reviewing source code, the more secure it is. The opposing side stated that closed source software was more secure since it was kept secret and only the developers were able to fix the code.

Today, that war has subsided with Microsoft, the biggest proponent of closed source development, beginning to offer the Linux operating system alongside Windows 10 and its server-based operating systems. With **Windows Subsystem for Linux** (**WSL**), you can install and use many of the Linux-based applications found on Red Hat Enterprise Linux or Ubuntu. Apple has also gone through a new phase of development by dropping its classic operating system for FreeBSD, and Linux has become the first operating system to be used on Mars!

Computer hardware has also undergone drastic changes – from the massively large computers found in the 1950s and 1960s to the Raspberry Pi of today, which is only a few inches long. Servers have transitioned from "big iron" systems that took up multiple units (or Us) of rack space to smaller one-U servers that contain more horsepower than computers built just a few years ago. We have also seen a large push to virtualization, having multiple standalone operating systems on one physical server.

We have seen a large push in server deployments as well. Although many organizations still deploy their applications on single, standalone servers, that is starting to change. Today, the containerization of application deployment has really taken off. This allows an organization to create single or multiple special-purpose operating systems that contain all the libraries and dependencies necessary to deploy an application.

Exploring server hardening steps

Servers and the data that resides on them are the lifeblood of many organizations. Banks use servers to allow their customers to manage their finances. Without servers, stock trading would be non-existent. Storefronts rely on many different types of servers and software architecture to sell products and services. They are used to receive information from IoT devices.

Businesses rely on information and without it, there is no business. That data can be in the form of sales information, intellectual property and trade secrets, information collected on a company's employees, emails, and many other forms. Safeguarding this information is crucial to the overall success of a business. If customers begin to lose faith in how a business protects their sensitive information, then that business will fail.

For many years, we used the network to compensate for deficiencies in software. Firewalls were created to block open ports or services running on the server. **Web application firewalls** (**WAFs**) came about to mitigate deficiencies in web application code. We micro-segmented the network to minimize the blast radius of a compromised server. Firewalls can also be configured to proxy web application traffic, minimizing your attack surface.

Moreover, recent attacks have shown that we can no longer just rely on the network to secure the environment. In 2016, the Mirai botnet began taking over insecure IoT devices that were connected to the internet. How did it do this? By attacking open Telnet ports with default credentials. It is estimated that the botnet spread to an estimated 600,000 connected devices used to attack websites such as `KrebsOnSecurity.com` and the popular DNS provider Dyn[1].

Hardening these devices, or at least changing the system defaults, could have prevented this botnet from growing to the size it did. The following sections will show you the various steps you can take to harden a device.

Operating system patching

We previously discussed how there are roughly 50 software bugs for every 1,000 lines of code. With operating systems now ranging anywhere from 50 million to 85 million lines of code, we are bound to have issues. The only way to mitigate vulnerabilities in the code is through patching. The early days of software patching were hit-and-miss. Patching software could bring down entire systems due to compatibility issues. Special environments were put in place just for testing patches developed by the manufacturer. This was used to test the patch first and if it worked, then deployment of that patch went to production.

For many, this was the course of doing business when patching. This also meant there was a substantial race from a vulnerability being made public, the software manufacturer releasing the patch, to getting to the point at which a company's IT department was able to implement it. Often, it took weeks if not months to install a patch, which meant that affected companies were vulnerable to whatever the patch was supposed to fix.

Microsoft has long released patches on the second Tuesday of the month, famously known as Patch Tuesday. While it made IT administrators happy to know when the patches would be released, many in the IT industry prefer rolling patches. Rolling patches often occur multiple times a month, resolving issues with libraries, services, and software. There really is no right or wrong way to do this, although having frequent patching is desirable. Operating system developers also perform out-of-band patching, or outside the usual patching cadence, when there is a serious vulnerability. This is all dependent on the manufacturer's security model.

Although less prevalent today, testing operating system patching is still the best practice. Microsoft, with good intentions, has released patches known to resolve many vulnerabilities – and inadvertently broken other services. For instance, in January 2022, Microsoft released Patch Tuesday fixes that broke the native operating system's VPN client. Many companies scrambled to remove the patch to get their IPsec and L2TP VPN connections back up and running. In August of the same year, Microsoft released a patch that inadvertently made Windows 11 computers unable to boot if their hard drive encryption software, BitLocker, was enabled. I may be picking on Microsoft, but Apple and other operating systems are not immune to these types of errors either.

IT should patch software and operating systems regularly. Now that Microsoft releases patches on Patch Tuesday, businesses have elected to patch their systems the following weekend. This allows patches to be tested before they are rolled out to the rest of the environment. Patching should also be part of your vulnerability management program.

Once a patch is released, IT or security should rescan the system to ensure that no new vulnerabilities have been introduced into the system. If any have been, you will need to work with IT to resolve whatever new vulnerabilities were introduced into the environment. Again, once the patch has resolved the vulnerability, you will need to rerun the vulnerability scan until the vulnerability has been removed from the environment.

Least privilege

Many of us, even today, still run our operating systems with full administrative rights. Proponents argue that running a system with administrative rights reduces the amount of help desk calls a company receives. Employees can install their applications for work or configure them for their needs. Others just want to make their company-owned **personal computer** (**PC**) their own by installing non-approved applications such as popular music software or video games.

There are plenty of operating system developers who have purposely removed or greatly reduced what the root or administrator user can do on a computer. Canonical's Ubuntu operating system, a Linux distribution, has disabled the root account altogether. Upon installation, the first user you create is provided *superuser* rights on the machine to perform administrative tasks. Apple's macOS, now loosely based on FreeBSD, has also disabled the root user account. These operating systems come close to implementing secured-by-default access, or least privilege, to the system. Microsoft, on the other hand, leaves securing user accounts up to the user.

BeyondTrust wrote an article in March of 2015 detailing how the reduction of administrative rights mitigates plenty of vulnerabilities. For instance, BeyondTrust states that 97% of all critical vulnerabilities would be mitigated if we ran our computers with reduced or user-level rights. With Internet Explorer, 99.5% of vulnerabilities and 95% of vulnerabilities related to Microsoft Office would be reduced[2] by running the operating system with reduced rights.

There are plenty of adversaries that rely on this type of setup as well. Malicious drive-by-downloads, macros in Microsoft Office documents, and even scripting embedded into PDF files all rely on the user having administrative access to their machine. When the macro or script is executed, it is executed with the same level of rights and privileges that the user logged into the machine has. This means that the executable that was downloaded without your knowledge can run on your operating system with full rights and privileges.

How can you go about reducing the number of rights a user has? Start with the least privilege principle. Least privilege is the idea of only providing the level of rights and privileges a user needs to perform their job function. If that level of privilege is not enough, then grant additional rights and privileges, but only to the degree required. If the user absolutely requires administrative access, consider providing that user with two separate accounts – one user-level account and another for administrative purposes. When trying to install, configure, or change a configuration file, that user will then be prompted to switch to their administrative account. Both of these accounts should reside in Active Directory or some type of **lightweight directory access protocol (LDAP)**. This provides a central way of auditing and logging all the accounts used in an environment. An example of how you could construct a user and administrative account is as follows:

- User account – `jsmith`
- Administrative account – `jsmith_admin`

The same holds for server administration. Do you absolutely need administrative rights when using SSH or **Remote Desktop Protocol (RDP)**? No, and it is poor practice. If a systems administrator requires elevated privileges, then they too should be using two separate accounts for remote management. Executing many different applications and services with administrative rights should be avoided at all costs.

Removing unneeded services

Years ago, operating systems came with all their services and features enabled by default. While this made it easy for administrative purposes, it greatly increased the attack surface. The more services that are installed and running on an IT resource, the more prevalent they are to attacks. Many administrators do not look at securing them either, leaving them running often without authentication.

If you are not using a service, turn it off or consider uninstalling it altogether. This will reduce the number of attacks against your systems and provide less administrative overhead. For example, you originally installed a web server that has a MySQL database. After a few months, you decide that you need to have a cluster of MySQL database servers for high availability and failover. Once the cluster is installed, you migrate the databases to the new cluster and point the web server to the new server. However, in doing all this work, you move on to the next project without removing the old MySQL service from the web server.

Days, months, and years go by and you forget that the database is there and has port `TCP:3306` open to the world. Although it may seem that this is a minor mistake, it could have harmful consequences. If that database is not monitored or secured properly, it could be open to attack by others.

Performing reoccurring network scans using NMAP can help identify forgotten and unneeded services. Once a service has been identified as no longer crucial to your business objectives, turn the service off and prevent it from automatically starting back up during a reboot. Vulnerability scans will help depict this as well. A vulnerability scanner will detect software running on the IT system and provide valuable information, including notifications about outdated software.

A periodic review of firewall rules is also needed to reduce your attack surface. In the previous example, in which we migrated the database to a MySQL cluster and there was no longer any reason to have that service open, you would need to remove the firewall rule that permitted other IT resources to connect to it. This will not only filter traffic from coming into that server but also help keep your firewall rules up to date.

Host-based security

For many systems administrators, the first thing you do after installing the operating system is turn off security features provided by the manufacturer. This includes host-based firewalls and many built-in security architectures such as **Security-Enhanced Linux** (**SELinux**) and **Application Armor** (**AppArmor**) used to protect the system from compromise. While this may make system administration easier, it also makes it easier for adversaries to take over IT resources.

SELinux was first developed by the **National Security Agency** (**NSA**) in the early 2000s and is still widely used today. It is a kernel module that provides **mandatory access control** (**MAC**) of the operating system. Among many different features, it labels files, directories, applications, and processes based on the following context: `user:role:type:level`.

When starting with development, you may want to set SELinux to permissive mode. This can be done in many different ways – for example, by either editing the `/etc/selinux/config` file or using the `setenforce` command. To review the labeling of the filesystem, use the `ls -Z` command. This will provide labeling information on everything in the directory. For instance, the Apache web server labels files using the following context:

```
unconfined_u:object_r:httpd_sys_content_t:s0
```

What happens if a file were to be moved from a user's home directory with the preceding context of `unconfined_u:object_r:user_home_t:s0` to the root directory of Apache? If SELinux is set to permissive, nothing. However, when you move a system into production, it should be set to enforcement mode. This blocks a file from running in that directory.

So, what is the point of all this? What if an adversary got hold of your web server and decided to place a web shell in the Apache root directory? With SELinux activated, the attack would have been stopped. Why? The context would be wrong for any files that reside in the directory. Since the context is wrong, this prevents the application from running. This has effectively stopped the adversary from taking over the system remotely.

SELinux is not foolproof, and it takes time to get used to running a system with it activated. There are ways to relabel the entire filesystem to ensure that the server will run as intended. The point here is to apply defense in depth to high-risk systems. Another tool that can be used is **AppArmor**.

AppArmor is also a kernel-level security module that can restrict how applications run on a Linux system. Much like SELinux, AppArmor relies on mandatory access controls rather than those based on user or group permissions. Ubuntu and many other flavors of Linux distributions rely on AppArmor to protect them from threats. AppArmor can easily be enabled on a system by running the `aa-enabled` command. To check and see whether AppArmor is enabled on the system, run `aa-status`. If AppArmor is running, you will see the following:

```
apparmor module is loaded.
40 profiles are loaded.
37 profiles are in enforce mode.
```

AppArmor configuration files are located in `/etc/apparmor.d` and you can view the configurations by using the `cat` command. Configuration files can be quite lengthy, and editing them is not for the faint of heart.

Server-based firewalls should also be enabled in a production setting regardless of using network-based firewalls. While network-based firewalls certainly have their place in security, host-based firewalls are also needed. In *Chapter 9*, we discussed how many organizations use fairly large LANs, which can sometimes accommodate 254 hosts. Once an attacker can gain access to a system, they can use that system to pivot their attacks against other systems in the network.

Host-based firewalls provide a stopgap, preventing an adversary from attacking adjacent systems. Applying rules to the host itself prevents many different types of attacks against services running on a system. Running NMAP against a system also provides valuable information on the type of operating system and the services running on it.

Picking the right password

One of the most important aspects of cybersecurity that you can teach your employees is to pick good passwords. Passwords can be tricky to create and are often hard to remember the longer they get. Long ago, it was acceptable not to have any passwords at all. Today, systems often require 12-16- character passwords. Bruce Schneier, a well-known cybersecurity researcher and board member of the **Electronic Frontier Foundation** (**EFF**), wrote a blog post on his research on how users pick passwords:

> *A typical password consists of a root plus an appendage. The root isn't necessarily a dictionary word, but it's usually something pronounceable. An appendage is either a suffix (90% of the time) or a prefix (10% of the time). One cracking program I saw started with a dictionary of about 1,000 common passwords, things like "letmein," "temp," "123456," and so on. Then it tested them each with about 100 common suffix appendages: "1," "4u," "69," "abc," "!," and so on. It recovered about a quarter of all passwords with just these 100,000 combinations.*

> *–Bruce Schneier*[3]

How many times have you had to call the help desk to get your password reset and the technician has picked something such as Summer2022!? If the mandatory password length is 10 characters and it must use a combination of uppercase, lowercase, numerical, and special characters, then there seems to be nothing wrong with this. However, there is plenty wrong with it. Commercial and open source password crackers such as John the Ripper and Hashcat could pick that password apart in seconds. What is also bad about this password scenario is that the help desk technician tells the user to reset it once they log in – but how many do? Users get busy and often forget to reset it, leaving the password in use for days or months.

Many organizations test their users' passwords on an annual basis. These systems can range anywhere from a small Raspberry Pi all the way to servers loaded with graphics cards to offload and utilize the processing power of **graphics processing units** (**GPUs**). John the Ripper and Hashcat, both open source and free to use, can be used to decrypt hashed database files such as Windows **Security Accounts Manager** (**SAM**) and Linux's passwd file. Password-cracking utilities are only as good as their dictionary files. Dictionary files are flat database files that contain a hashed password, along with its clear text password. For instance, if we were to use the previous example, Summer2022!, as our password, the SHA1 (explained in more detail in the *Encryption* section) hash would look like the following:

```
ef9a6f5bf9f36b2e2487f0b174990a581ca8c044
```

Reusing passwords across multiple systems is also a bad habit to get into. Social media sites such as LinkedIn and Facebook have had their password databases stolen and put up for sale on the dark web. Facebook had its user password database stolen in 2019, with millions of passwords stored in clear text. Many users, once they pick a password, think that it is secure and will reuse that password over and over again, putting your company's information at risk.

Users should get used to the idea that they cannot necessarily trust the security of another system. They cannot perform audits against Facebook's password storage, for instance. This is why using a password manager can be ideal for your organization.

Using password managers

Many password managers perform zero-knowledge encryption on the password database. This means that the contents of the password database are never known to the service provider. How do they do it? They use a single, master password to store all of the sensitive information. That database is then either stored in the cloud or locally on the user's computer. The master password is then only known to the user and never shared with the service provider. That way, the service provider only has access to the encrypted blob stored in their cloud.

There are many benefits of using a password manager. One, you could in theory never know what your password was for a given website, as the password manager manages all of that for you. Two, you must – and I emphasize *must* – use a secure password that only the user knows. Why? The database used to store all the password is only as secure as the master password – so if you use Summer2022! as your master password, game over. Many password managers offer **multi-factor authentication** (**MFA**), which is an additional way of protecting your user account from adversaries.

How do you pick a secure password? There are a thousand theories on how to do this and if you ask two people what they do, you will more than likely get different answers. Many use passphrases, taking sentences that are meaningful and modifying them to make them hard to guess. For instance, if you were to use "*My daughter's name is Samantha*," you could make a passphrase such as `MyDaught3rsN@me!sSam`. This passphrase is long, fairly complex, and easy to remember.

Adding randomness to a stored password

Salting passwords has grown in popularity over the years and for good reason. Anyone can hash a password and use it against a password database. That hashed password is common across any system too. For instance, if we were to hash `Summer2022!` on one system, it would be the same hash on another system. This gets service providers into trouble if their password database were stolen. To protect against this, I recommend adding a little bit of salt to the database.

Salting a password does not mean pouring table salt on the server, although that would be hilarious! Salting is a method of adding padding that is not known to the user to a password. Salt can range anywhere from 4 to 6 characters in length, although it can be longer, randomized, and applied at the beginning or end of the password hash. For instance, if our salt was `a1b2` (please do not use that) and we wanted to apply that to the end of the password string, it would look like `Summer2022!a1b2`.

`a1b2` is applied automatically without the users' knowledge. That way, when you store the password in the database, it strengthens the password further. For instance, the hashed value from our example password was the following:

```
ef9a6f5bf9f36b2e2487f0b174990a581ca8c044
```

Now, by adding a salt value to the password, it becomes the following:

```
6fa6a9357ff4e420901df4da69d4de188ca88704
```

If companies were to salt their passwords, no two password hashes would be the same across service providers of the same password. This would not eliminate the issue of reusing the same password across multiple systems – however, it would make it difficult for adversaries to gain true knowledge of the password itself.

Using break-glass passwords

Break-glass passwords are also important. They are passwords that are generally used by local accounts or those that are not tied to an LDAP or Microsoft's **Active Directory Directory Services (AD DS)**. These accounts can include system passwords such as the administrator or root accounts of a system. They can also include database passwords, such as the `sa` password for Microsoft's SQL Server.

While picking passwords is one thing, you should only use break-glass passwords if and when an IT resource is unresponsive. Typically, the passwords to these accounts are either stored in a secured password vault or written down on a piece of paper. No, I am not saying that the password is written on a sticky note and placed under your keyboard. These passwords are typically written down on a piece of paper, secured in an envelope, and stored in a safe. Along the crease of the envelope, you can either write the name of the person who changed the password or the name of the system on which the password is being used. This protects the envelope from being opened and then closed again without breaking the seal.

MFA

Passwords were great and had their place in time. Today, passwords are still important, but we have fallen victim to poor password hygiene. As mentioned earlier, we tend to use simplistic passwords that are easy to crack and reuse passwords across multiple accounts to easily remember what was used. We continuously see **account takeovers (ATOs)** from brute-force attacks while others use passwords discovered on the dark web. How can you ensure your users are who they say they are when accessing your corporate systems and data?

The importance of using **MFA**, or **two-factor authentication** (**2FA**), provides an additional layer of assurance. This additional layer forces your corporate users to use another factor besides their username and password. It requires the user to present themselves in a way that proves they are who they say they are – but how?

MFA builds upon common usernames and passwords. It requires the user to perform an additional authentication that is outside of something that they know. For instance, 2FA requires two of the following three:

- **Something you know**: *Something you know* essentially is the username and some type of secret. This secret can be in the form of numeric and alphanumeric characters. These characters are then randomized (or not) in such a manner that the password is not easily guessable. **Personal identification numbers** (**PINs**) are also a type of secret, although they are less secure as they typically only require numeric characters.

- **Something you have**: *Something you have* builds upon the username and password combination by forcing the user to have access to a physical device during authentication. That physical device can then provide a randomized string of digits that must be typed into the application. Other methods include receiving a push notification to a pre-registered cell phone, text messaging, and callbacks. Many out there discourage the use of text-based messaging, as that can also be attacked. While this is true, I would argue that text-based MFA is still better than no MFA at all.

  Something you have can also include a physical key that is plugged into a USB slot or operates via **near-field communication** (**NFC**). The U2F specification was developed by Yubico and Google. This standard allows the user to use a physical key that is registered with a service provider as the secondary factor. U2F works by entering in your credentials first, as you normally would, and then tapping a physical key connected to your computer, typically a key inserted into the USB of a computer system.

- **Something you are**: *Something you are* is not widely used as it can be costly and there are privacy concerns about using it. However, the use of retinal scanners, fingerprint readers, hand scanners, and voice recognition can all be used for this. In the U.S., many states have enacted privacy policies to reduce the impact on individuals who use the technology. For instance, the state of Illinois created 740 ILCS 14, a bill aimed at protecting state residents from misuse of their sensitive data.

Secure software configurations

It does not matter whether it is closed source or open source, software in general is insecure. That goes for operating systems, databases, web servers, and even Microsoft Office. It is deployed that way to make it easier for the consumer to utilize it for their own purposes. Software can be deployed with default credentials, insecure default encryption, and active guest accounts, even with missing patches. It is up to you and your team to ensure that it is as reasonably secure as possible before deployment.

You can search for default credentials online for just about any software. This makes it easy for an attacker to gain access to your systems. Password crackers and brute-force software make it almost foolproof to use, and script kiddies are out there every day scanning systems across the internet. Before deploying a system into production, make sure that any default passwords are changed. Systems are developed with default accounts too. Occasionally, guest accounts are created so that others can gain access to a system. These should either be secured or disabled.

Apache with OpenSSL and even Microsoft's **Internet Information Services (IIS)** are deployed with default encryption standards. In many instances, these are deployed with insecure encryption algorithms that allow attackers to take advantage of exploits such as POODLE or HeartBleed. It is up to you and your team to ensure that the encryption being used is sufficient to today's standards. Need help? SSL Labs[4], provided by Qualys, will scan a website and gather information about the secure algorithms being used. Once the scan is complete, it will provide a grade based on the encryption levels being used.

As an example, I used a quick scan using `sslyze`, an open source SSL/TLS scanning utility, against a default install of **Red Hat Enterprise Linux (RHEL)** 9 and Debian 11, both on their latest release. While the default installation of RHEL blocked my attempts to connect over TLS for TLS 1.2 and lower, 1.2 was the only one to respond. The default Debian installation shows that TLS 1.0 and above were able to handle connections. This is poor practice, as HeartBleed is susceptible to versions 1.0 and 1.1 of TLS.

There are also plenty of security benchmarks that can be used to secure an application. The Center for Internet Security provides documentation on how to secure many of today's popular software and operating systems. For instance, you can gain access to many of their benchmarks for free, which include how to harden Microsoft and Linux systems, cloud-based software, databases, DNSes, firewalls, load balancers, and web browsers. The U.S. Department of Defense also provides similar benchmarks called **Security Technical Implementation Guides (STIGs)**[5].

STIGs are like the Center for Internet Security benchmarks in that they provide an overview of how to securely configure an IT resource. To use the STIGs, you must have Java installed on the machine. After that has been installed, you need to download the STIG viewer as required XML files for the resource that you intend to secure.

Changing system defaults

System defaults… every operating system and hardware has them. Ever tried to log into a router without knowing the default password? Try Googling *Default Linksys password* and instantly you will receive page results providing default credentials for you to log in. You might receive default credentials such as *admin* and *password* or, in other instances, it could be *Cisco* and *cisco*. My favorite default credentials are *root* and *calvin*. Why? I used to read Calvin and Hobbs growing up.

Default passwords are not the only configuration item that should be changed. SNMP also comes with default read-only and read/write strings, those being public and private, respectively. What is the danger in this? Quite a significant one. A quick search on `shodan.io` provides some scary results. Approximately 21 million devices connected to the internet have SNMP exposed to the internet. Many of them have never been reconfigured, allowing the default strings to be used.

SNMP is used to monitor device statistics such as the network bandwidth or hard drive usage. SNMP also allows you to change configurations of the device through its read/write string. This would allow anyone to change the configuration of a device remotely from across the internet. Changing system defaults is so important that many regulatory bodies require organizations to modify their systems before deploying them to production.

Remote management

How many of us work outside of the office? Better yet, how many of us work outside of the data center? I am guessing pretty much all of us. We as IT professionals perform most if not all of our system administration remotely. This means we rely heavily on remote administrative services such as Windows RDP, SSH, and even lights out management such as the **Integrated Dell Remote Access Controller (iDRAC)**.

Remote management utilities were not developed with security in mind. Systems then were still very primitive without a lot of processing power, which encryption heavily relies on. Remote management utilities such as Telnet and **Virtual Network Computing (VNC)** protocols were used as the standard for remote management. While the use of these protocols would be frowned upon today due to being clear text, you still might see them in production today.

Legacy network equipment is still prominent in organizational networks. Why? Organizations do not have the **capital expenditure (CapEx)** to maintain 5 to over 1,000 pieces of network equipment that range in cost anywhere from $4,000 to $8,000. If you wanted access to a GUI such as the Windows desktop, many used the VNC protocol for this, as it made systems troubleshooting easier.

Performing remote administration using one of the protocols mentioned meant that all your usernames, passwords, and even the screens you looked at on the desktop could be decoded using packet capture utilities such as WireShark. *Sniffing* the network allowed even more insight into the communications of the systems. Websites used the clear text **hypertext transport protocol (HTTP)** for many years. If someone were able to perform a packet capture against the network, that person could grab websites, voice calls through the **voice over IP (VoIP)** protocol, and even bank account information. The industry needed a change.

As the processing power of systems began to mature, we started to see encryption being used. This brought encryption to newer authentication methods such as SSH. SSH is now the standard for use when accessing a **command-line interface (CLI)** used by Linux/UNIX systems and networking equipment such as Cisco or Juniper. It wraps the communication going back and forth between a user and the IT resource being administered. As with most authentication protocols, it uses asymmetric encryption, creating a private-public key pair. This allows the administrator to not only use a username and password for authentication but also use SSH key pairs for authentication, removing the ability to authenticate using a password, which is more secure.

There are a few ways to protect yourself if you still have clear text protocols in use. First and foremost, you want to terminate the encrypted tunnel as closely as possible to the endpoint you are trying to reach. For instance, we could place a VPN device as close to the IT resource as possible. To use Telnet or other clear text protocols, you first must connect to the VPN, and then connect over clear text to the endpoint. To restrict the use of Telnet from across the network, you would need to place a filtering device, such as a firewall, between the IT resource and the rest of the environment. You could add the following to your network to secure clear text protocols better:

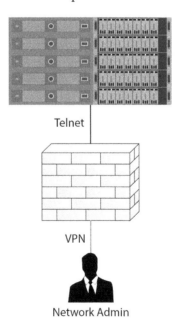

Figure 10.1 – Remote administration with Telnet

While the previous example is the most secure, it is also the hardest to set up. It requires re-architecting the environment by placing VPN appliances in one or multiple places throughout the network. Many clear text protocols now offer the ability to be wrapped by TLS, offering the ability to add encryption to a protocol that previously did not support it. Protocols such as RDP, VNC, and database protocols allow this. However, it requires an organization to set up a **public key infrastructure (PKI)** to generate certificates.

There are quite a few things that you can do to better protect your infrastructure from being attacked. The no-cost or low-cost solutions discussed in this section, when implemented, make it harder for an adversary to take over an IT resource. The next section will discuss warning labels you receive when connecting to an unverified device.

Zero trust

Zero trust is a concept in which you no longer trust any device connected to the network, even to your internal LAN. Many of us have employees who are given laptops and mobile devices to use in the office and at home or a coffee shop. They are also given administrative rights to their devices to install any software package they want. This lessens the overall security posture of the device.

The only way to trust a device connecting to your network is through a series of checks of not only the user but the device itself. These checks include MFA for user authentication and software checks ensuring the device has the latest updates and virus definitions, and you may also want to check that the device does not have certain software installed, such as BitTorrent, Bitcoin mining, music streaming software, or video games.

Another thing to consider when implementing zero trust is understanding the device connecting to the network. Is it domain-joined? Is the device managed through InTune or JAMF? Are there attributes of the device that we can check to ensure that it is a company-owned device? These could include checking whether a company-deployed certificate is on the device.

Zero trust is a concept that you can use to protect your environment better – it is not a technology that you can buy. The U.S. federal government is spending a lot of time, money, and effort putting this into place. For more information, follow the link in the *References* section – NIST SP 800-207[6].

Dangers of TOFU

We ran into a huge problem when we started implementing SSL on websites. How can we trust the identity of the website we are accessing? End users were told as long as there was a padlock on a website, they were safe. They blindly access a website and immediately trust it even though an error message on the browser saying **This site cannot be trusted** appears – but it has a padlock, so it must be okay, right? Unsuspectingly, a user clicks on **Take me to the site** and enters their credentials, only to find out later that it was a phishing website performing credential harvesting.

End users are just doing what they have been told to do, looking for the padlock to ensure their information was safe. They want to gain access to information or rectify whatever problem there was on their account as quickly as possible. We are all guilty of it and still do it to this day. As an IT or cybersecurity professional, how many times do you initiate an RDP session only to receive an error saying that the remote computer you are accessing cannot be verified? How about trying to connect to an IT resource through SSH only to get an error message that the authenticity of the remote host cannot be established?

By blindly accepting these error messages, we too are developing bad habits. We **trust on first use** (**TOFU**) any remote connection we make, as we assume devices are to be trusted. How can we be sure that this is one of our devices? How do we establish trust from across the LAN to another IT resource?

When software is installed for the first time, it generates encryption keys and certificates. These keys and certificates are self-signed, meaning we have not established trust with the device. First, we must establish trust with a **certificate authority** (**CA**). By establishing trust, our end user devices will validate and accept certificates signed by our CA. However, we must install and configure our own internal CA to do this. This can be done in several different ways. If you are a Microsoft shop, the simplest and easiest way of doing this would be to install Microsoft's Active Directory Certificate Services. Do not fret if money is tight – you can still create your own with open source software. Install any flavor of Linux you prefer and use OpenSSL.

We can do the same with SSH and a CA. The CA will be used to sign both the user and host keys. By signing both the user and host keys, we establish mutual trust between the two. We can then perform granular role-based access by setting access permission, even only to a few hosts, to expire. Performing SSH key signing prevents us from blindly trusting a user or host key. Anyone could drop an SSH key on a server and have it work. Signing the keys prevents that from happening. Plenty of medium to large organizations have put this in place to further secure their deployments. For more information, please check out the writeup on Facebook's engineering page[7].

The IoT

When **Internet Protocol version 6** (**IPv6**) was developed, it utilized a completely different format for addressing. Typical IPv4 addressing comes in four octets separated by three periods (such as `192.168.0.1`). However, IPv6 is quite different. Instead of numerical representations of the address, we use hexadecimals with 8-bit octets separated by a colon. A few organizations have become more clever in the way they obtained their addressing. For instance, Facebook's IPv6 address is `2a03:2880:f172:81:face:b00c:0:25de`.

Did you see `face` and `b00c` in the address? There is even a dead beef address, `FD00:DEAD:-BEEF:64:34::/80`.

With IPv4, we have roughly 4 billion addresses available for our use. As mentioned in *Chapter 9*, we discussed how we have run out of IPv4 addresses. Network and port-based address translations have saved us, but for how long? The term IoT was coined in 1999. Today, it seems to be what everyone is talking about – the ability to make any device internet-capable. This will drive the adoption of IPv6 faster than we had expected. As 4 billion just does not seem to be enough for everything, how many devices can IPv6 handle? Well, if we were to give every grain of sand an address, we still would not run out. That is 2^{64} hosts for the entire internet.

Making any and every device internet-capable? Yes. Lightbulbs, toothbrushes, health trackers, smart home devices, locks, garage doors, and even heart monitors. Vehicles are now equipped with monitoring devices used to assist with getting directions or calling medics if you were in a car crash. Most of these devices have IP addresses too, as they all talk over IP and as such require an address. As with everything else, these devices must be secured, but how?

IoT devices are not typical computers. Most are considered embedded devices that have software running on them and are internet-capable. Software patching is not as simple as running Windows Update or `sudo apt dist-upgrade`. No, in the early days, these devices typically required special software or hardware to be connected to the device. Most consumer-based IoT devices can now be upgraded over Bluetooth or automatically updated over the internet.

How can we be sure that updates come from the manufacturer? How do we know that we have not missed an entire update? IoT devices do not have a ton of memory or hard drive space, but they now play a critical role in our daily lives. Through the use of encryption, we can verify that the software being installed on the device is correct. Let's dive into that next.

Understanding encryption

You may think of encryption as a way for bad people to keep information secret from authorities. This battle is nothing new – it has been around since the 1990s and continues today. Encryption, however, has many benefits, as you may or may not know. For instance, we can ensure documents have not been altered, validate software, verify someone's email address, and even ensure the privacy of our data.

The simplest benefit of the bunch is validating documents using hashing. When we spoke about passwords in this chapter, we learned that hashing is used to protect a password from prying eyes. A hash is a one-way encryption method that can be used to encrypt passwords; however, it can also be used to verify the authenticity of a document. What if you wanted to create a document and ensure that it was never altered? A hash of the document is one way of doing this.

If you were to create a file with `Hello, world` written in it, you would get a SHA1 hash value as follows:

```
7b4758d4baa20873585b9597c7cb9ace2d690ab8
```

Now, change the words to `Hello, world!`, and its SHA1 hash value becomes the following:

```
09fac8dbfd27bd9b4d23a00eb648aa751789536d
```

As you can see, the hash values are not even close. We can now tell that the document was altered in transmission between being sent by the writer and received by a recipient. We can do the same thing with software.

While not as common, software developers now sometimes provide hash values for their code. Software supply chain issues have become a problem in recent years and consumers want to ensure the software they use is legitimate. Providing a hash is one way of ensuring that the software is unaltered. What if we also wanted to ensure that it came from a trusted source?

Digital signatures

Digital code signing certificates are another way of ensuring that code is unaltered. The benefit of digital code signing is the ability to present a public certificate, signed by a trusted third party, proving ownership of the code. Anyone can provide a hash for the code being developed. However, when an organization goes through the process of digitally signing the code, it says that they were the ones who developed it.

When digitally signing code, you must create a private key first. This private key should be kept, well, private. No one other than the organization that will be using it should have access to that key. Once the process of generating a private key has been completed, we then create a **certificate signing request** (**CSR**). The CSR is then presented to a CA to be signed off. Before getting the CSR signed, the third party who will be signing the request will perform a validation of your company.

This validation often requires proof that the organization exists, its location or address, and often requires signatures from the CEO. This validation process is needed to truly verify the existence of the organization. Once validation is complete, we now have the CSR signed. Once the CSR has been signed, we will then receive a public certificate. Public certificates are in an X.509 version 3 format that contains all the attributes necessary to validate the organization and the certificate. For instance, if you were to go to Amazon's website, you would be able to see the attributes of the TLS certificate being presented to you. This is the same premise for a certificate being used to sign code.

Apple and Google require software to be digitally signed for it to be publishable in their app stores. Microsoft requires kernel driver modules, among other operating system software, to be digitally signed. Apple's macOS also requires that software be digitally signed before it can be installed. If the software is not signed, it will either present an error message stating the software is not signed or provide alternative ways of installing it by skipping validation checks. This can be dangerous, as you cannot trust or verify the integrity of the software.

Protecting the private key

When generating private keys in software, anyone with server access, in theory, has access to the private key. The reason for this is that the private key was generated in this software and stored on the server's hard drive. This poses yet another problem. Server administrators, or anyone who had access to the server for that matter, also have access to the private key. If an adversary were to gain access to that key, they could essentially decrypt anything that was encrypted using that private key. Even worse, an adversary could digitally sign software using your private key, and offer it to the masses.

This can break the separation of duties between server administrators and those who are responsible for ensuring the security of cryptography. We want to ensure the integrity and security of the private keys being generated so that they do not fall into the wrong hands. It could spell disaster for an organization if those keys were ever lost or stolen.

Cryptographic keys can be used for many different applications. While we tend to think of encryption as being used for protecting our data when accessing a website, encryption is used for many different things. It can be used to protect sensitive configuration information or used in the operations of a system. In some instances, such as the healthcare or automotive industries, it could result in hundreds of thousands if not millions of dollars being lost due to recalls or lawsuits.

To combat this problem, special hardware has been developed to protect a private key from being lost or stolen on the server. **Hardware security modules (HSMs)** provide an additional layer of trust for protecting the private key. With HSMs, it is a best practice to generate the private key in hardware. This typically means better entropy for the key being generated. Second, when creating private keys in hardware, once generated, they cannot be extracted. Preventing the key from being extracted from the HSM means that the key cannot be stolen by an adversary who has access to the server.

There are many HSM manufacturers out there willing to sell you an HSM for thousands, if not hundreds of thousands, of dollars. For small to medium organizations, this places the security of your keys being protected way out of your price range. HSM manufacturers such as NitroKey[8] and Yubico[9] provide HSMs for a reasonable price and work with software such as Active Directory Certificate Services or OpenSSL.

Encryption plays a key part when protecting documents and software. It can prove who you are and show whether a document was altered. The protection of the private key is important. If it were ever lost or stolen, it could be used against you.

Summary

It is all about the data; without it, there would be no business. Computers hold onto and allow users and administrators to manipulate that data when needed. It is up to you to safeguard the data from those who want to harm your organization. The steps outlined in this chapter, such as patching software, changing defaults, picking the right passwords, using MFA, and planning for system downtime, are critical for your business to function.

Encryption, too, has many different applications, whether safeguarding passwords or ensuring a document was unaltered. As we've discussed, itz is also used to protect the privacy and security of remote authentication to outside resources. The issue though is that we must first establish the trust of the issuing CA. Without establishing trust, we cannot validate a secured remote connection.

Following these few steps can help assure that your systems are protected against most threats. Nothing is foolproof and results may vary by organization. However, a reduction in administrative rights, ongoing patching, and changing system defaults will protect you from threats against your organization.

In the remaining chapters, we will focus on secure application development and how to build meaningful metrics that can be used to show your executive team and the board.

References

1. *Inside the infamous Mirai IoT Botnet: A Retrospective Analysis*: `https://blog.cloudflare.com/inside-mirai-the-infamous-iot-botnet-a-retrospective-analysis/`

2. *Removing admin rights mitigates 97% of critical Microsoft vulnerabilities*: `https://www.beyondtrust.com/blog/entry/removing-admin-rights-mitigates-97-of-critical-microsoft-vulnerabilities`

3. Schneier, Bruce. *Passwords are not broken, but how we choose them sure is*: `https://www.theguardian.com/technology/2008/nov/13/internet-passwords`

4. SSL Labs: `https://www.ssllabs.com/`

5. DISA STIGS: `https://public.cyber.mil/stigs/`

6. NIST SP 800-207 – *Zero Trust Architecture*: `https://csrc.nist.gov/publications/detail/sp/800-207/final`

7. *Scalable and secure access with SSH*: `https://engineering.fb.com/2016/09/12/security/scalable-and-secure-access-with-ssh/`

8. NitroKey: `https://www.nitrokey.com/`

9. Yubico: `https://www.yubico.com/`

11
Securing Software Development through DevSecOps

Historically, security in the **software development life cycle** (SDLC) has not received as much attention as it should. We have built a defensive network security architecture to combat poor software design. Legacy systems that were introduced years ago, sometimes decades ago, with little to no documentation leave IT and security teams scrambling when a new software vulnerability is discovered. Supply chain issues have crept up in recent years, sabotaging legitimate software vendors and open source contributors.

The term **DevOps**, or **development and operations**, has been around for a long time. The term stems from the combination of these two teams working collaboratively. The software development team needs IT and operations in order to deploy their code to production. There are reasons for this, but it is mainly to maintain separation of duties. Many organizations, even today, do not provide this separation. This often leaves IT administrators confused as the development team pushes code without their knowledge, and if something breaks, it then becomes IT's problem to resolve. If IT was not involved, it is a safe bet that security was also not involved either.

Security was often a second thought when it came to software development. Developers were rushed to push code for new features by hard deadlines and could not be slowed down. As corporations began to get attacked, this mindset shifted – to the left. Shifting left in the life cycle meant that security teams would be brought in early in the planning and development phase. When we add security to the DevOps process, we call it **DevSecOps**. Even with bringing in security teams to evaluate the code, we still see a fairly large number of vulnerabilities. As we learned in the last chapter, there is an average of 50 software development issues per 1,000 lines of code.

We have had to change our methodology in project management when it comes to coding too. Old project management methods, especially around the Waterfall approach, left making changes to code a nightmare. This often doomed any changes to the code, including security changes. This meant that changes or implementation could not be performed in a timely manner. There had to be a new way of coding that included operations and security teams.

Throughout this chapter, we will focus on software development, focusing on how you, as the head of security, can make security a priority in the SDLC. We will review testing methodologies, along with how and when to use them and when to make changes to the code to ensure secure code development.

This chapter will discuss the following topics:

- Why introduce cybersecurity early?
- A new style in project management
- The six principles of DevSecOps
- Code reviews
- Open source licensing
- Gitflow branching
- Secure coding checklists
- Embedded secrets

Why introduce cybersecurity early?

There are plenty of ways to make you and your team successful when it comes to cybersecurity. In *Chapter 1*, we discussed a few steps you need to take within the first 90 days, one of those being forging friendships. Well, not so much friendships, though many friendships grow out of the workplace. It is more of a matter of introducing yourself to your team, department, and company. You need to make relationship building a priority so that others can reach out to you and feel comfortable doing so. Relationship building is truly needed in IT, especially in cybersecurity as we tend to be known as the department of "No!"

Now that you have built those relationships, it is time to be integrated into the cogs of the business. This means that you will need to work with all sorts of departments to help them understand that security is a necessity. This often means that you will need to be a key process-changer for many of the IT processes found within the organization. This means, with the assistance of others, you will need to understand the process and how security and the rest of the organization can work together to achieve the end goal.

Traditional software development methodologies were not meant to change the development life cycle. This often meant that changes to the code, no matter the stage, could not be changed until the end. This posed a problem in many different areas. One of these was that, due to miscommunication, what the customer or business wanted and how the development team interpreted it could be two completely different things. Programs were written without any testing or input and the outcome was not known between all parties until the very end.

While these methods got the code out the door quickly, it left the stakeholders wondering why they received an application that did *X* when they really wanted something else. This also meant that testing, that is, security testing, was not built into the process and often was not pulled into the project until the very end. If testing was not built into the SDLC at the very beginning, it could mean disaster for all those involved.

The term *shifting left* was brought about to help change this mindset. Shifting left means that security must be thought of at the very beginning and used throughout the SDLC. Continuous testing is needed to efficiently build security into development so that changes can be made during the process. This provides less waste: waste of time, waste of money, and so on. How can you do this when development teams are nose down, pounding away at the keyboard?

A new style in project management

Agile allows for testing and changes to occur during the life cycle. While Waterfall could last from a few weeks to several months, Agile allows the team to break up the implementation of features or tweaks in the code into multiple chunks through the use of sprints. Sprints are very short, often only lasting for 2 weeks, allowing software development teams to make changes and push code.

As stated, sprints allow development teams to break up coding projects into small chunks or units of work. During the development process, the teams work off a backlog of work that is pulled forward to be worked on next. There are plenty of reasons why teams work on certain projects. They could be following a last in, first out policy. This means that new features are added to the bottom of the backlog and will be worked on when that feature bubbles to the top. Another reason is because of priority. This means that no matter what is in the backlog, these features take priority for being worked on over everything else in the backlog.

This also allows for security testing to occur on a regular basis. As code is being written, it can be continuously checked to ensure that it meets the company's security standards. Security issues that were discovered could either be worked on prior to being pushed to production or placed in the backlog to be completed at a later date. Without Agile, this would not have been possible.

Agile project management has its roots in the Agile Manifesto. The 12 principles of the manifesto are centered around communication between team members and the continuous delivery of software to the customer. While Agile is mainly centered around software development, it can also be used in other types of IT- and security-related projects.

While Agile is centered around project management, DevSecOps also has a set of principles. The six principles of DevSecOps can be integrated into Agile project management to properly plan out and execute software development for your organization.

The six principles of DevSecOps

As previously mentioned, DevOps was primarily driven by development and operations, leaving security out of the life cycle. Much like how development and operations must integrate and work together, we cannot forget about security. DevSecOps is a newer method of integrating all the key players. DevSecOps is built on six principles all strung together to make testing, especially security testing, an automated process. The six principles are as follows:

- Plan

- Build

- Test

- Deploy

- Operate

- Monitor

Let's look at each of these in much more detail in the next few sections.

Planning

While we want to automate as much as possible, there are certain processes that require more of a personal touch. The planning stage is just that. When you are planning for your new software application, you go through the process of identifying the problem. Rarely do we jump in and start coding right away. There needs to be an outline of what we want to do and how we will possibly achieve our result. The planning phase should also include timelines or milestones of when we plan to have a finished product. Planning also includes security. This needs to be outlined to understand who, what, where, when, and how security will be woven into the project plan.

We typically think of applying security to the coding process through scanning the written code, running applications, reviewing open source libraries, and even thinking about the licensing of the code we use. This is also the time to begin creating our **systems security plan** (**SSP**), which will evaluate the security posture and register risks of the application, who will be using that application, and its stakeholders. In addition to that, we should also include threat modeling to understand the various risks the application can and will have once it is placed into production.

To review these risks, we will use the **STRIDE** model, which stands for **spoofing, tampering, repudiation, information disclosure, denial of service, and elevation of privileges**:

- **Spoofing**: Typically, we think of spoofing as an internet problem. It is the ability to fake one's IP or email address to impersonate something or someone. Spoofing can include other things as well, such as illegitimate programs masquerading as the real thing. It can also include websites that are stood up for phishing or stealing credentials from your users.

- **Tampering**: Tampering can be done by legitimate and illegitimate people. Accidents happen, and everyone is guilty of it. However, how easy it is for someone to tamper with data should be reviewed. Does someone without administrative rights have the ability to change key parameters of the system or the data? If legitimate users can easily do this, how easy would it be for someone who is not supposed to have access to the data? Security safeguards should be in place to prevent changing, or make it extremely difficult to change, data that the user is not supposed to have access to.

- **Repudiation**: Ever received an email from someone only for them to say, "I never sent that!" This is an example of repudiation. Repudiation is when one needs to prove that something happened, even if someone or something says otherwise. Often, this requires log files to be written once on an append-only filesystem. This assures that when a log event is captured by the system, it can never be altered by anyone. Digital signatures on emails sent from individuals are another way of knowing that the email came from the sender and was not spoofed by a third party.

- **Information disclosure**: Information is most important to any organization. **Intellectual property** (**IP**), or an organization's secret information, must be kept private and they should only allow those who need to view the data. Information disclosure is when the organization accidentally permits documents to be viewed when they were not supposed to. This also includes adversaries who want to gain access to confidential information. The organization must keep that data secret and not allow it to fall into the wrong hands.

- **Denial of service**: Availability, part of the CIA triad (along with confidentiality and integrity), must be reviewed as part of the overall system architecture. If an adversary were to knock the system offline, that would be considered a denial of service, preventing others from accessing the system. Denial of service has played a large part in security and IT in general. It sometimes consists of sending various types of data to an IT resource, overwhelming it to the point where it can no longer function. Denial of service can occur either internally or externally depending on the system's location.

- **Elevation of privilege**: While spoofing your account can elevate your privilege, depending on the account, the elevation of privilege looks at how someone can promote themselves to an administrator using a non-administrative account. Elevation of privilege is the ability to manipulate a system or application to provide the user with more rights than what they should have. This can be detrimental to a system as it can then allow the user to change system parameters or look at sensitive information that they would otherwise not have access to.

Once we have completed the planning stage, the next thing is to begin building the application.

Building

As the planning phase comes to a close, we begin to build the application. Developers begin coding to build out certain functions and features of the application. Some developers may incorporate open source code or libraries to help build that functionality. During the build phase, we need to continuously check the code and libraries to ensure that coding mistakes have not occurred and the libraries are free of vulnerabilities.

One way of detecting mishaps is to use **static application security testing** (**SAST**) tools to determine the overall security of the application. The SAST tool should be used during the build process to perform the scans. If any issues arise during the scans, they should be remediated and reevaluated to ensure that the mistakes and vulnerabilities have been mitigated.

Testing

Once we have an application created, it is time to move it to a test environment. The test environment should be an isolated, internal-only environment where we can perform various tests against a running application. These tests include user experience and quality assurance, but they also should require security testing. As the application is in a running state, we cannot use SAST to scan the application as that relies on static code.

Dynamic application security testing (**DAST**) tools are used at this stage as they can be used against a running application. Some **version control system** (**VCS**) providers can perform these types of scans against your code as you commit them to your organization's repository. If you do not have this ability, you can use other tools, such as OWASP ZAP[1] or PortSwigger's Burp Suite[2]. Both tools allow free use, while Burp Suite also provides a commercial offering of their tool.

Deploying

Once we have performed all the required testing, and source code errors and vulnerabilities have been fixed, it is time to roll the application to production. Before we do this, however, we must perform one important step, official sign-off. The SSP provides valuable information about the application and overall architecture of the new system being placed into production. It is your responsibility to ensure that the risks have been identified and mitigated prior to going live.

Operating

At this stage in the game, we can sit back with our favorite beverage in hand because the work is… never done! At this point, the application is running in a production environment, and it is our responsibility now to ensure that the lights stay on. This does not mean that the work is done. We need to ensure that the application is up and running at the expected response and availability times. This can mean the use of load balancers and scaling the application to ensure the application is always up and running. This also ensures that the application and the IT resources are backed up properly.

Monitoring

Monitoring can seem like the easiest thing we can do, but it is not that simple. How do we know that we are monitoring the right things? What are the key metrics we need to capture and report on? Are there new vulnerabilities to the application or underlying libraries? We need to monitor all of it and then some. This is to ensure that the application is running as expected and that we can mitigate any vulnerabilities that have been detected.

Automated ways of performing code scanning to look for vulnerabilities in your code should just be the start. As your team grows and matures, manual code reviews should also be part of your overall review process. You should build the team to understand and perform both. Defense in-depth does not stop at the network; it should also be introduced into your SDLC too.

Code reviews

Performing code reviews can be one of the best ways to detect errors in the code base, if they are done properly. All too often I have seen great development teams produce excellent code only to fail on the reviews. Failing does not mean that the review was not happening; it should happen at every merge or **pull request** (**PR**). The failure was actually reviewing the code. Many times, throughout the day, I would hear, "Can someone please do a code review?" only to hear 5 seconds later, "All done!" What was the point in even doing a review?

Manual code reviews are not something that is really taught in college, at least not when I was there. In fact, secure coding was not something that was taught at all. We cannot blame our developers for not doing something that they were never taught. We have to teach our developers what to look for when they are performing the reviews and how to correctly modify code to make it secure.

There are a few different ways to promote good coding hygiene. Make sure that the developers are properly trained. First and foremost, ensure that the developers understand the code they are reviewing. If the code was written in Java, ensure that a Java developer is reviewing the code. Code reviews should also be performed by mid- to senior-level developers in your organization. They are better equipped to understand the code and perform the reviews than, say, a junior-level developer. This does not mean, however, that the senior developers should not have their code reviewed also; they should.

Build metrics around performing the reviews. How many reviews did a developer perform in a given day or week? How long did it take the reviewer and how many mistakes were corrected? While metrics may seem hard to grab, stats are provided through many different VCSs in use today. Data from these metrics should not negatively affect the team; this should be used as a tool to baseline how well your team is performing.

Should code reviews be performed at every branch of code that a developer is working on? It depends on how you want to standardize when these should occur. Code reviews must occur at any point in time when a developer is merging code into the main branch. Where in the branching of your code you enforce reviews is up to you, wherever you feel most comfortable. However, again, they must be placed into some type of standard and be reviewed by all developers, so they too understand when to do them and why this is important.

SAST, while it can be performed as the developer is writing code, can also dramatically help others who are required to review the code.

Static application security testing

As we begin to look at the various tools that allow us to perform security testing, we first need to understand what testing looks like. For that, we initially turn to SAST, which is probably the easiest, cheapest, and quickest way to perform security testing as it looks at static code rather than running code. Static code is code that is not running on a system.

Programmers are paid to write code – obviously. Unless you are a hardcore programmer who uses Vim to write your code, chances are your developers use an **integrated development environment** (**IDE**). There are plenty of IDEs out there to choose from, with many IDEs being written for a specific language. For instance, Eclipse is used for many different languages, but it is primarily known by Java programmers. PyCharm? That would be for Python development. C or C++ programmers use **Visual Studio** (**VS**) if they are in a Microsoft environment. VS Code is becoming a Swiss Army knife because of its ability to work with many different languages and being cross-platform.

There are a few different ways to utilize SAST during the development life cycle. One way is to allow the programmer to write code and perform security checks at the end. The SAST tool allows the programmer to upload their code to a repository where it is scanned for coding mistakes. Once those mistakes are identified, it allows the developer to go back and fix those mistakes prior to going to production or wait for the next sprint.

SAST can also be incorporated during the development process through an IDE. While the programmer is writing code, the IDE can perform these checks before the code is committed to a VCS such as Git. This provides continuous feedback and allows the programmer to write and fix the code in real time, rather than waiting to upload it to the scanner itself.

Ease of use is your first step toward adoption by developers to become security focused. I have seen organizations fail at getting this right, but it was not for a lack of trying. Your first step is to get adoption from the executive team. Once this is done, the next step is to get application developers to use it. Too often, developers are given strict deadlines they have to meet in order to get the product out the door. They cannot be bothered with developing, then compiling, then uploading, then waiting for test results – you get it.

When SAST is done correctly, you will get adoption from all corners of the organization. It has to be easy to use and the test results have to be easily understood. Vulnerabilities discovered during the scanning process must be reviewed and risks mitigated. Without the adoption of the tools, this will not happen. Development teams will also not use the tools if they are hard to use, often requiring training of development staff in order to effectively use the tools.

An additional benefit is that SAST analysis tools are more efficient than manual code reviews. While code reviews are still necessary, they can be performed less frequently than performing a scan using a SAST tool. One thing to note is that there is no silver bullet when it comes to security and automated tools are not the guardian angel we are looking for. However, when layering SAST and manual code analysis together, you are one step closer to achieving the security posture that you are looking for.

Dynamic application security testing

SAST scanning is considered *white-box* testing as it needs access to the source code in order to perform the scans. As mentioned previously, it is also the easiest, and also considered the cheapest, scanning option available. To get a complete picture of the application, you need to scan it multiple times in various different ways. How can you be sure that once the code is compiled and running there are no new vulnerabilities discovered?

Unlike SAST tools, DAST requires a running application to scan. DAST tools are also considered *black-box* testing as the system testing does not need to see the underlying source code to perform security checks. DAST checks are also not run as often as SAST. This is because not all applications are ready to be scanned in a running environment.

During the development stage, the application may not be at a point where it is ready to be placed into an environment where it can be run. This can be true for code that needs to be compiled correctly, such as Java. Other languages that rely on hypertext markup, such as HTML or PHP, do not require being compiled for them to run, making it much easier.

Though DAST is not used as often as a SAST tool during the development stage, it should be used regularly post-deployment. Vulnerabilities change constantly and it is up to you and your team to ensure that the applications are secured against them. Dynamic scanning is one way to ensure that the application is somewhat free of those vulnerabilities. PCI DSS, for instance, requires that applications that store, process, or transfer credit card information must be scanned on a quarterly basis. I challenge that this is not frequent enough.

If you rely on the scan reports to be quarterly, how will you know if new vulnerabilities exist for the web app? An example would be if you had an application vulnerable to Log4j. Corporations were scrambling to discover whether their application was vulnerable. If you waited to scan the application for another 2 months and the application was vulnerable, you would have no way of knowing until it was compromised. Scans should be performed at least monthly, if not sooner.

Many cloud-based VCS providers either offer DAST tools as part of your subscription, or they can be integrated using outside tools such as OWASP's ZAP[1]. Typically, these tools are utilized as part of the build pipeline. This is to ensure that the application is secured appropriately prior to moving the application to production. Once the application is in production, you need to resort to scanning on a regular basis to ensure the security of the application and its data.

We also need to test and verify the open source licensing and applications being used in our application. Software composition analysis assists with that review.

Software composition analysis

In December 2021, it seemed like the internet was on fire. Security experts, bloggers, and even news outlets were warning anyone and everyone. IT and security teams needed to figure out whether a little-known open source library called Log4j was running on their systems. Vulnerabilities of this nature were nothing new to security experts. With a CVSS rating of 10 out of 10, organizations needed to patch it, and fast.

Log4j is used to capture logs of the underlying application. It can also be used to log memory usage of the system, error messages, user access, and so on. Log4j is used in just about every Java application developed by small companies to large IT and security vendors.

The problem was that people did not know whether they were susceptible to it. You see, IT professionals did not have insight into the code or libraries being used by their vendors, making the problem worse. **Software composition analysis (SCA)** can help with this.

SCA is used to scan developed code and their associated libraries, much like SAST does, but provides different results. An SCA scanner will dive deep into the code base and determine what open source components are being used and create a report showing the details. The report will also show any vulnerabilities that were discovered in open source libraries so you can make risk-based decisions on whether to upgrade to the latest release.

Another key component of this is the ability to generate a **software bill of materials (SBOM)**. Much like a normal **bill of materials (BOM)**, an SBOM is used to depict all the software components used in the application. This becomes critical when trying to respond to an issue such as the one found in Log4j, as you can now pinpoint which applications are susceptible and can resolve the problem quickly, or at least get prepared for when a patch is developed so that you can implement it quickly.

For example, say you are writing a Python script to interact with a MySQL database. To do this, you import a number of different modules, including the MySQL module, into the script for you to use. Once it's ready to be packaged, you create the `requirements.txt` file, which shows all the different modules and the version number. At the time, at least you hope, the entire package is free of coding mistakes and vulnerabilities. Now, if you are a consumer of that script, how would you know? By using SCA, you get a picture of the risks associated with using that script and the necessary modules that are needed to use it properly.

As an organization, you must also review the various types of software licenses being used in your open source software. Many software licenses require that closed-source applications also include open source code to be provided to the public. This could introduce problems with your business objectives. As you will discover in the next section, Cisco was sued for not releasing portions of its source code to the public that were originally licensed under the GPL.

Open source licensing

Closed source software can make it extremely difficult, if not impossible, to reuse parts of their code for other use. In many cases, organizations consider the source code proprietary or intellectual property, not something to be shared with the outside world. There are two sides to the open versus closed source debate and both sides state that security is the overall reason for the release of developed code.

Those on the closed source side say that the reason for not releasing source code is because of security. If no one outside the company can see the code, then no one will be able to review it to see vulnerabilities. This debate has gone on for many years. Today, many still view closed source as a good security measure. If an organization can keep their secrets close to their chest, then adversaries cannot exploit security holes in the application.

The argument for the open source debate says just the opposite. If the code is open source, then anyone is able to find vulnerabilities in the code base. If there are hundreds of thousands of people looking at the code, then vulnerabilities are easily detected. Oh, and if someone were to find such an exploit, they have the ability to submit code to patch said vulnerability. This means that patches are developed and applied quicker than in closed source as more people are viewing it.

Neither side is right or wrong. However, there have been some issues that arose in the past with open source code. Take, for example, Red Hat, a commercial Linux-based open source operating system (that was a mouthful). As it is open source, the source code had to be released the public. For instance, their source **Red Hat Package Manager** (**RPMs**) software can be found on Red Hat's FTP site (`ftp.redhat.com`). This has made it possible for other companies to copy and redistribute the **Red Hat Enterprise Linux** (**RHEL**) operating system. As an example, Oracle has taken the RHEL source code and created their own Linux operating system called Unbreakable Linux.

This is nothing new. The second oldest Linux distribution, Debian, has also been replicated multiple times. It is also the father (or mother – I guess it depends on how you look at it) of Canonical's Ubuntu. Ubuntu has been spun off to other popular Linux distributions, such as Mint and Pop!_OS. Imitation is the sincerest form of flattery, I guess? All joking aside, you can see how open source has opened up fierce competition against many different software vendors.

With these issues, why use or create open source software? In the Python example, open source software packages can be used in many different applications, making it easier for the developer. Closed source is not immune to using open source either. In 2003, Cisco Systems was sued by the **Free Software Foundation** (**FSF**) for using open source software in their Linksys products. The lawsuit stated that the Linksys routers used open source code, which was distributed under the **General Public License** (**GPL**). This allowed Cisco to use open source code; however, it required the company to publish the modified code to the public.

As you can see, not all open source software is created equal. In the next section, we will discuss the different types of open source software licenses and how you can use them in your own products.

There are many different types of copyleft and permissive licenses that open source software can be distributed under. As the head of security, you should familiarize yourself with the various types of licenses open source software can be distributed under. It is also the time to include your legal department or outside counsel to review what copyleft licenses you should use with your company's software.

Copyright licenses

Traditionally, copyright licenses are used to protect one's work from reproduction. Book authors, for instance, use copyright to prevent someone else from reproducing or taking credit for their work. Computer manufacturers use copyright to protect ideas that may or may not be used in upcoming products. Software vendors, much like those who produce closed-source applications, use copyright to protect their code from being reproduced or copied in a competitor's product.

Copyright does not stop there; you can even copyright functionality. For instance, in 2018 the dating application Tinder sued Bumble, also a dating application, for stealing IP. Tinder alleged that Bumble stole their ideas when it came to swiping left or swiping right when looking at someone's profile.

Linux has also been targeted for supposed violations. In the early 2000s, Microsoft began to sue large Linux distributions claiming that it violated 235 patents. Microsoft also claimed that users of the open source software were also required to pay a fee in order to use a distribution of linux or be sued. At the time, Novell, the parent company of SUSE Linux, agreed that it would pay damages to Microsoft.

One of the benefits of using open source software is the ability to share ideas through code. To some, code is considered like poetry to be shared with everyone. The use of open source licenses allows for the sharing of code while maintaining certain protections from use by others. These licenses are called copyleft.

Copyleft licenses

Copyleft licenses allow the use, modification, and distribution of work provided by others. The caveat to this, however, is that redistribution of work must accompany the same rights and privileges allowed by the developer. This also means that software or works created under a copyleft license cannot be changed to a copyright license.

For example, say your company creates a software program and distributes that work under a copyleft license. A user or organization would be in violation of the original license if they were to redistribute the same or modified software under a copyright license. In other words, software distributed under a copyleft license must provide the same rights and privileges as the original creator of the work. Written by Richard Stallman of the FSF in 1989, the GNU GPL is what most people think of when it comes to open source licenses.

GNU General Public License (GPL)

The GNU GPL was the first and most widely used open source license that an individual or corporation could adopt. It provided the freedom to allow anyone to use, modify, and distribute their code without restriction. If an application contains any code that falls under the GPL, the entire application the developer wrote now falls under the GPL. The GPL does not enforce the free distribution of software. code can also be obtained through commercial means or by paying a fee. However, it must be freely distributed, without fear of use or modification. Free in this context means freedom of use, not free as in monetary value.

There are three different versions of the GPL license. First released in 1989, GPL version 1 was the work of Richard Stallman. Stallman, a computer programmer, has developed a number of popular applications still in use today, such as GNU Emacs. Stallman is also the founder of the FSF, used to promote open source software.

GPL version 1 was created as a way to merge a few different licenses as Stallman's works were licensed individually. This first version was also created to stop proprietary software developers from releasing binaries and other executables without ways of being able to read the source code. The other issue was that corporations, who might use or fork open source code, might use it for their own gain and make the once open source code closed source.

GPL version 2 was created in 1991 as a way to prevent corporations from using open source code and making it proprietary, that is closed source, regardless of their previous obligations. This meant a corporation could not take open source code, use it in their software, and resell it to the masses, regardless of other legal obligations the company may have.

The GPL version 2 license is still used in many mainstream applications today. For instance, the popular **content management system** (**CMS**) application WordPress utilizes GPL version 2. The CMS software has been estimated to be installed for millions of websites worldwide. These include the White House website (`whitehouse.gov`), CNN, UPS, and eBay.

GPL version 3 was created in 2007 and is used in large software projects, such as the Linux kernel. It allows anyone to take and modify the code as they saw fit; however, that version of the code is subjected to the same rights and obligations as the original code. This means that the modified code also has to be distributed under GPL version 3.

Permissive licenses

Permissive licenses are quite different from copyleft licenses. If distributed under a permissive license, anyone, including other organizations, can utilize the code without restrictions. This includes code integrated into proprietary software that an organization does not want to redistribute. Let's look at a few types of permissive licenses used today.

MIT license

The **Massachusetts Institute of Technology** (**MIT**) has created an open source license that allows a third party to use any or all software created under it to be integrated into other software. This includes proprietary code developed under a copyright license. Anyone who obtains code under the MIT license can do what they want with it; however, they must include the original copyright and license in future copies of the software or source code. Others cannot hold the original developer liable for any damages that may occur while using the software.

Apache license

Much like the MIT license, those who wish to use code or software developed under the Apache license can do what they want with the code with few exceptions. The Apache license is quite detailed in comparison to the MIT license, providing additional protections that must be included. This also means that the protections one might receive in the MIT license is not the same as one would receive in the Apache license. These additional protections are as follows:

- Original copyright notice
- A copy of the license itself
- A change log of any modifications made to the original code

The change shall detail the modifications made; however, the developer does not have to detail their IP. In addition to the benefits of using the Apache license, many developers and organizations choose this license because it is backed by the Apache Software Foundation.

BSD 3-clause license

The first version of the BSD license was created in 1980 and was used to cover the **Berkeley Software Distribution operating system** (**BSD OS**). Similar to the MIT and Apache licenses, there is very little in the way of restrictions when it comes to its use. Companies are allowed to use BSD-licensed code in their proprietary applications. Those who intend to use code under this license are required to include the following:

- The full text of the license
- The original copyright notice

As previously mentioned, there are stark differences when it comes to copyleft and permissive licenses. Policies and standards should be created to ensure that the licenses being used or consumed are within the risks and business objectives that align with the company. They also have to align with the goals that you and the organization set and how you distribute the code. Is it, and always will be, closed source code? Make sure that your developers do not include code or libraries that use the GPL.

Discovering what licenses, code, and libraries your developers are using can also be time consuming. This is why I encourage anyone using open source, or may unintentionally use open source code, to scan it with an SCA tool. Not only will this build out the SBOM, but it will also help you discover the type of license and its vulnerabilities.

Gitflow branching

How often do your developers store code on their local hard drives? I will bet that if you do not standardize on a VCS or source code repository, your IP is not protected. A VCS is not only meant to store your code; it also allows others within your organization to share and modify the code as needed. This collaboration is a necessity when dealing with large groups of developers all working on the same code base.

There are many different types of VCSs out there to choose from, but one of the most popular is **Git**. Linus Torvalds, the creator of the Linux kernel, is also the founder of Git. Git allows developers to commit their code to a centralized repository. A repository is a place to store, almost like a backup, code for current or later use.

Git also allows you to pull historical information about the code being committed to the repository. You can see past commits, when they were committed to the repository, who did it, and the logs of each transaction. When done correctly, external third parties can also audit the repository to ensure that processes are being followed properly.

One of the more powerful features of using the tool is branching. This allows your developers to work on specific tasks or features in the code without hindering other developers who are also working with the code. In essence, you could have one, two, or even hundreds of developers all working on the same code base, but one group is working on features while another group is working on hotfixes. Gitflow branching has five different branch types. Let's look at each of them in detail:

- **Main branch**: Each repository starts off with a main branch. Consider this as production-ready code for your application. As such, the main branch must be protected as much as possible. The main branch is the first branch each repository starts with. When committing to the main branch, code reviews must occur at every PR and the commit should be protected and digitally signed. The digital signatures prove that one of the company's developers worked on and committed the code to the repository. This prevents anyone from merging unsuspected code from another machine in the event of an account takeover of that user's account. This also helps with software supply chain issues.

- **Develop branch**: The develop branch holds onto all your preproduction code that is developed. All feature branches should be created from the develop branch. Once the feature is complete, that code should be merged back into the develop branch for additional testing and review.

- **Feature branch**: The feature branch is where you add new functionality to your code base. As previously stated, all features should be created off of the develop branch.

- **Release branch**: The release branch is where you prepare and plan for new production releases. Release branches could include security or bug fixes that were discovered during the testing phase and need to be integrated into the code prior to its release to production.

- **Hotfix branch**: Hotfixes include any required changes to the main branch. These are used to fix or patch any issues discovered in the main branch. Once a patch has been created, it must be committed back to both the main and develop branches. This is to ensure that the hotfix has been placed in both branches and it is not missed.

Introducing Git and the suggested workflow for branching promotes best practices for how you should work with the code you and your team write. All of this does not matter, though, if you are not working with secure coding practices. The following section provides a few places to review when creating secure coding practices.

Secure coding checklists

As mentioned previously, secure coding can be difficult to achieve as it is not often taught in schools. Many developers, if they have the time, research secure coding practices either at work or while they are at home in their free time. Resources have been created to assist with the methods of secure coding; the problem is just finding the best method for your needs.

Three organizations stand out in providing secure coding practices: **NIST**, the **Software Engineering Institute** (**SEI**) at Carnegie Mellon University, and **OWASP**. Both organizations provide material and checklists for how to train your employees and what to look for when evaluating code.

NIST

With assistance from BSA, OWASP, and SAFECode, NIST has developed the **Secure Software Development Framework** (**SSDF**). The framework laid out in SP 800-218 assists organizations in establishing a method for maintaining software throughout the SDLC. The criteria for the SSDF are as follows:

- Prepare the organization

- Protect the software

- Produce well-secured software

- Respond to vulnerabilities

The criteria laid out by NIST circles around, first establishing a need for the organization, then providing awareness and training for why secure software development is required for organizations to produce good code. This all starts at the top level of the organization beginning with the executive team. There need to be protection processes in place to ensure that the code that is produced is protected from adversaries who wish to do harm to your organization and your customers. The last two center around secure coding practices and how the organization responds to and mitigates vulnerabilities discovered in your code.

SEI

SEI[3] provides secure coding standards when it comes to popular coding languages. They have developed documents on the following:

- C
- C++
- Android
- Java
- Perl

Their documentation is broken into languages, rules, and recommendations. As you continue to drill down into the documentation, SEI provides coding examples of what should and should not be in your code. If you are looking for assistance for another programming language, SEI has also created a checklist of things to look out for when writing your code. These include the following:

- Impose input validation
- Heed compiler warnings
- Architect and design for security policies
- Keep it simple
- Deny by default
- Use least privileges
- Sanitize data sent to other systems
- Program with defense in-depth methodologies
- Ensure quality assurance
- Adopt a secure coding standard

OWASP

OWASP, the organization behind the OWASP Top 10, has also created a document promoting secure coding practices. The OWASP Secure Coding Practices Quick Reference Guide[4] provides a checklist for the following secure coding practices (link in the *References* section):

- Input validation
- Output encoding
- Authentication and password management
- Session management
- Access control
- Cryptographic practices
- Error handling and logging
- Data protection
- Communication security
- System configuration
- Database security
- File management
- Memory management
- General coding practices

While the document was released in November 2010, it still discusses many of the pitfalls we see in computer programming. OWASP has additional secure coding and threat modeling tools available on their GitHub site as well.

Embedded secrets

In October 2022, Toyota was involved in a security incident that resulted in 300,000 email addresses being exposed. Was this due to poor password hygiene? No. Was it due to a SQL injection? No. So how did the adversaries get access to so many email addresses?

It involved a developer who mistakenly set one of the company's GitHub repositories to public. This allowed attackers to review the source code of a project that also happened to have an access key embedded in the code. This allowed anyone to siphon the email addresses from Toyota's T-Connect website.

Embedding any secrets, whether they are usernames and passwords, API keys, or access tokens, is bad practice. You should discuss this with your company's development team to ensure this is not occurring in your environment. One way to do this is to use AWS Secrets Manager. Secrets Manager allows the developer to grab a secret through the AWS API and utilize the key at runtime.

Other tools, such as GitGuardian, can scan code in a VCS to detect embedded secrets and private keys stored in code. Once detected, you can take remediation steps to remove the sensitive code from the repositories.

The point here is that you can feel safe that your developers are not embedding secrets or private keys in open repositories. This can better protect you and your organization from other attack vectors that may impact your organization.

Summary

Traditional information security had us using the network to protect our applications. Today's threats are more sophisticated, directly targeting our web applications. Sure, we can apply network controls to our applications, but they aren't enough when it comes to building out your information security program.

Newer processes have allowed us to begin the discussion of how we build security into the application pipeline. In this chapter, you have learned how DevSecOps processes, along with Agile methodologies, can allow for information security to perform testing on web applications before it is production ready. Code reviews, SAST, and DAST scanning allow us to scan the application and discover vulnerabilities and coding mistakes. SCA tools help us discover the types of open source libraries being used and their potential licenses.

Additionally, we learned about VCSs which are designed to allow developers to work on code collaboratively. This allows multiple developers to work on the same code base at the same time. We also looked at gitflow, one of the many methodologies to use when creating branches of features and hotfixes.

In the next chapter, we will discuss how to create good metrics and what to report to the executives and the board.

References

1. OWASP ZAP: `https://owasp.org/www-project-zap/`
2. Burp Suite: `https://portswigger.net/burp`
3. SEI: `https://wiki.sei.cmu.edu/confluence/display/seccode/Top+10+Secure+Coding+Practices`
4. OWASP Secure Coding Practices Quick Reference Guide: `https://owasp.org/www-pdf-archive/OWASP_SCP_Quick_Reference_Guide_v2.pdf`
5. OWASP's GitHub page: `https://github.com/OWASP`

12

Testing Your Security and Building Metrics

We have covered a lot of different topics throughout this book. We have built our vision and mission statements, aligned our security program to a framework, performed assessments, created a strategy, and learned about several different security topics. So, how do we know everything is working as intended? We must set up monitoring tools to ensure that our IT resources have stayed within our security baselines.

How do we do this? First, we build meaningful metrics to ensure that we are not only maintaining our security but are also mitigating the risks in our organization. Vulnerability scans, phishing attempts, ping sweeps, and brute force attempts are all things that we can report. However, we must build a story behind these in order for them to make sense to the executive team and the **board of directors** (BoD).

We are not only grabbing metrics from IT resources; we must also build metrics around projects and processes. Metrics should include testing intervals for how often you perform a tabletop exercise or test a mitigating control. What were the results of your red team/blue team exercise? How did the test of your incident response plan go, and how often are you doing it? How often do you review the organization's policies, standards, and procedures?

In this chapter, we'll be covering the following topics:

- Business and regulatory requirements
- Maintaining corporate and third-party security
- Why CARE about metrics?
- Vulnerability metrics
- Incident reporting
- System security plans and the risk register
- Risk heat map and balanced scorecard

Understanding your requirements

As the head of security, it is your responsibility to understand the distinct types of requirements imposed on the organization. These include, at a minimum, business and regulatory requirements. Let's gain a thorough understanding of what these requirements are.

Business requirements

By now, I hope you have picked up on a theme. IT and information security policies and processes all roll back up to business requirements. It is up to you, as the head of security, to take the business requirements and make sense of what is needed. Does the business want to increase its security posture? What does that mean, and what does that entail? How do we test our program to ensure it meets business objectives?

You must also create your own set of requirements for how you want to evaluate your security posture. The business is not going to tell you whether scanning your systems every 4 months is either too long or too short. Those decisions will be left up to you and your team to decide. You must, however, decide how often you want a particular control evaluated.

As mentioned previously, security is not a set it and forget it type of methodology – it has to be evaluated regularly. IT and information security change rapidly, and the controls that you put in a week ago may no longer work in today's environment. You must set a given interval for what you want to evaluate and the types of controls you want to report on.

How you evaluate your security program may differ depending on the market your organization is in. For instance, if your organization is strictly U.S.-based, then you do not have to worry about regulatory requirements from other countries, such as the **General Data Protection Regulation** (**GDPR**). If your organization is geographically dispersed, then you must worry about different response times for incidents, where you can and cannot store sensitive information, and different privacy laws. This can allow you to build different metrics depending on the number of **data subject access requests** (**DSARs**) you receive in a day, month, or year, for example.

Metrics not only help you justify adding more employees or deciding whether or not a control is meeting your requirements, but it also helps build a story for your program. This story can be used in many different applications, such as presenting to the executive team or the BoD. Metrics are meaningless until you have developed a story to highlight your accomplishments. Historical data also helps in the creation of that story. Did the number of DSARs increase month over month? If it did, why? How is the overall risk posture of your security program decreasing? What mitigations were put in place that created such a decrease?

Regulatory requirements

Not all regulatory requirements are the same, as we have previously discussed. Many regulatory bodies require annual testing of your security program, whereas others require more frequent testing. For instance, the **Payment Card Industry Data Security Standard** (**PCI DSS**) requires you to submit an attestation or have an outside auditing firm come in and evaluate your information security program annually.

If you have e-commerce websites that take credit card information for purchases made online, you must also scan your websites with a valid tool on an **approved scanning vendor** (**ASV**) list. These companies resell their products in order to perform various types of scans and have been vetted by the PCI council to ensure they meet their requirements. This does not mean that you can submit an Nmap scan and call it good. You need to partner with a third-party company in order to perform these scans.

Other examples, such as a **service organization controls** (**SOC**) report, are only valid for 12 months and are performed by an external third party, which is also a licensed CPA firm. While you can be assessed as many times as you want, if the report goes beyond 12 months, it is no longer considered valid. Other audits, such as those found in the ISO 27000 series, are good for 3 years. These endeavors are not something that you can get overnight, however. The ISO 27001/27002 assessments can take anywhere from 6 months to 18 months, depending on how quickly you want to get assessed and how much money you are willing to spend. I have seen some assessments with very short turnaround times, upward of $125,000 or more.

Regulatory bodies do not do this to make money, though it does cost a lot to get and maintain your certification. They want to ensure that when they assess an organization, their clients meet or exceed their security requirements. If an organization fails to meet the certification body's requirements, it are not only doing a disservice to its client but also to anyone who wants to do business with that client.

What happens if you fall outside of their requirements? Plain and simple, you lose your accreditation or compliance certificate. Many organizations utilize SOC 2 Type II reports and ISO 27001/27002 certifications to market themselves. These are meant to be handed out to other third parties, telling them that your security program meets the minimum requirements to achieve that certificate. If you do not have or do not maintain your certification, that could mean that you lose business due to non-compliance.

There are plenty of organizations that will only do business with you if you have one of these certifications. They trust you to keep their data safe and out of harm's way. While the media may mention your company in a breach report, it is their neck on the line to ensure their data is kept as secure as possible. While these are internal initiatives, they could mean a loss of revenue if you do not maintain a certain level of security and privacy.

Building metrics around internal security can be a lot of effort and will take time. If you are just starting out, you may not have historical data to show how the program has progressed. Everyone has to start somewhere; however, it is important that you begin collecting data as soon as possible to help build that story of what you are doing and why it is important.

At times, business and regulatory requirements may conflict with each other. For instance, you may have a business requirement that states you scan your system for vulnerabilities every month, whereas a regulatory requirement may want you to scan a system every week. When evaluating what is required, you must stick with the most stringent control necessary to meet an objective. You will need to adjust your program to ensure you meet all the objectives, not just the ones that you decide.

Continuous evaluation of your security program

You will want to continuously evaluate your security program against the strategy you laid out earlier. Building metrics around this will show how your program has evolved over time. This will come through internal and external evaluations of the program. When maintaining your corporate security, metrics will show how the security program has evolved and *hopefully* increased over time. Performing this for your trusted third parties will also hold them accountable for their security program.

Maintaining corporate security

Continuous evaluation of your security controls is going to play a big part in maintaining your corporate information security program. You need to re-evaluate your security controls and how you are aligning with your security strategy and developing **key performance indicators** (**KPIs**). You must also take into account how your security plan still aligns with business objectives.

Did you set out a series of goals that you wanted to achieve this year? Did you meet your intended target of reducing risk to your environment? How would you know unless you are measuring KPIs? You must continually measure yourself against previous metrics to get a viewpoint of how you have progressed over time. For example, you use an attack surface management tool to evaluate your external threat sources. At a given interval, you receive reports that state what risks you have exposed to the internet, and those values go up or down depending on what is exposed. As you take action in removing or reducing the amount of risk being exposed, those reports begin to look favorable. This would be a great KPI to report on.

KPIs do not have to be technical in nature either, they can also be for administrative controls. An example of this would be how many policies, standards, and procedures were developed over the past year. How did that proactively affect your alignment with the NIST Cybersecurity Framework? Did it elevate your maturity score? Using documents such as the NIST CSF scoring template[1], you can see how the current state of your information security program has progressed over time. It is recommended that you at least evaluate yourself against the CSF annually to ensure that you are staying on the right track.

Another way of measuring the success of your security program is through the use of **objectives and key results (OKRs)**. OKRs are about goal setting, meaning they represent what you plan to achieve during a particular time period. While an objective can be high-level, the key results should be specific. OKRs are also quantitative in nature, depicting whether or not the goal was achieved in the time allowed. An example of an OKR you could use is set out here:

- **Objective**: Reduce vulnerabilities in IT resources:

  - **Key Result #1**: Purchase vulnerability management software

  - **Key Result #2**: Scan 100% of the environment

  - **Key Result #3**: Decrease overall vulnerabilities by 25% in Q1

The time frame for when you want to achieve a goal objective is up to you. Many organizations create OKRs for the upcoming year as a way of project planning.

I would also look into other free services, such as the **Center for Internet Security Controls Self Assessment Tool (CIS CSAT)**. The CSAT is a free tool that you can use to measure yourself using the CIS benchmarks. Once you have evaluated your organization, the CSAT will provide reports that can be used for BoD presentations or as a summary report for your controls. You can also use it to assign tasks to others in your organization.

Maintaining third-party security

So, you have made sure that your company's information security program is being maintained. How do you ensure that the companies you do business with are doing the same? You have trusted third-party vendors with your data, it is your right to ensure that they are maintaining that security to your business and regulatory requirements. Many companies have added a *right to audit* clause in their contracts with other companies.

The right to audit is not something new; however, it is rarely used in contracts. Many organizations rely on third-party assessments and certifications, such as the NIST CSF or ISO 27001, respectively. Some third-party regulations, such as PCI DSS, are point-in-time assessments of the organization's security posture, while others are ongoing audits. While asking for these documents up front is one thing you should do, you should also be allowed to independently assess your third-party security controls.

Now, you are not going to have the ability to perform assessments on everyone. Large companies such as Amazon, Google, and Microsoft will not allow third parties to assess their internal security controls. In these instances, you may be able to sign a **non-disclosure agreement (NDA)**, which will allow you to view more sensitive control documents, such as a SOC 2 Type II report. Many large organizations will also post what certifications they have. For example, AWS has posted its ISO certifications on its website for public consumption. This is used to provide comfort that external third-party auditors have evaluated the security of AWS, and you can trust it with your sensitive data. If you want detailed information as part of your overall review, you may be able to, but most will only provide one with an NDA in place.

In addition to evaluating third-party information security, you should evaluate fourth parties as well. Fourth parties are trusted vendors or suppliers, anyone who is doing business with your trusted third party. For instance, you and your organization have chosen GitLab as your preferred VCS. According to DNS, GitLab is using Google Workspace for its business email. Your company, however, is using Microsoft 365. In this example, GitLab is the trusted third party of your company and Google is a fourth party to you. You might ask "How is that?" Your organization has an indirect tie to Google because you do business with GitLab. If Google Workspace experienced a security incident, because you have provided sensitive or intellectual information to GitLab in the form of emails and source code, that incident could directly impact you and your organization.

There are a few paid-for service providers that will evaluate third- and fourth-party security on your behalf. Companies such as BitSight[2] and Security Scorecard[3] scan the internet, looking for various signals they can report on. These signals can be misconfigurations, open clear text ports, and outdated certificates, among others. These ratings are metrics that can be used to evaluate not only your security posture but the security posture of your third and fourth parties too. Your organization may already have a security score. As these service providers scan the public internet, any publicly facing assets you have are also being scanned. Many will provide a free ratings report for you based on the information they currently have obtained.

These service providers gather public information through regional internet registries such as **American Registry for Internet Numbers** (**ARIN**) or the **Asia Pacific Network Information Centre** (**APNIC**) to pinpoint IP address allocations and understand company-owned assets through common names in digital certificates or through names in DNS. They use this information to correlate assets to your company. Based on the assets they believe you own and the signals the service provider gathered, they generate a score for your organization.

Security ratings help paint a picture of the progress of your security program. They can be used to show how you and your organization have increased or decreased the security posture over time. They will also provide similar scores of your third and fourth parties as well. This is to help the executive team determine what risks they are willing to accept. If they are willing to only place **personally identifiable information** (**PII**) or other sensitive data such as **intellectual property** (**IP**) in a company that scores a B or higher, then you may want to look at your current contracts and perform an evaluation.

If you have a small to a mid-sized organization that you do business with, they may allow you to perform an audit. That third party may charge you billable hours to perform the assessment; nothing is free. However, it provides you and your organization with a comfort level when you better understand the third party's risk and security posture.

In *Chapter 3*, you were introduced to the NIST **Cybersecurity Framework** (**CSF**). The CSF is a great way to not only introduce a framework to the third-party company if they are not already using one but also to score their information security maturity. By using the scoring sheet introduced in *Chapter 3*, you have a simple yet easy way to score yourself or another organization's current state.

Using security ratings is an easy way to show how your organization and your third parties are aligning with security objectives. So, how do you go about developing metrics for your internal controls? One way is by using the CARE method.

Why CARE about metrics?

In 2021, Gartner released a new information security framework that is used to evaluate an organization's credibility and defensibility of an information security program called CARE. CARE stands for the following:

- **Consistency**: This is used to evaluate whether your controls are implemented properly over time. You will need to periodically perform this evaluation to ensure that the controls implemented 6 months ago still perform the same as they do today. If the controls are not working as intended, you will have to go back and re-evaluate them. This is in line with the **plan, do, check, act** (**PDCA**) approach we introduced in *Chapter 2*. We plan for the implementation of controls, complete them, check to see whether they are working as intended, and if not, decide what to do next.

- **Adequate**: Do the implemented controls align with business objectives? Do they meet or exceed expectations from stakeholders such as your executive team and the BoD? This will determine whether the controls that you and your team have implemented are adequately protecting the environment. Metrics will have to show whether your spam filter is blocking malicious emails or how often an antivirus database is being updated.

- **Reasonable**: This is used to measure whether controls are appropriate for the business. How do they align with the organization's overall risk appetite? Do the controls negatively impact confidentiality, integrity, and availability? If a system is secured but no one can access it, is it reasonably secured? To some security professionals, I am sure the answer would be yes. However, if no one can access it, then what is the point?

- **Effective**: The effectiveness of the control must also be evaluated to ensure the correct outcomes from the implemented control. Are you and your team mitigating threats effectively? Did you accidentally leave an AWS S3 bucket open to the public internet? How are you evaluating the systems in your environment to ensure the controls are effective? Again, these are all metrics that you should report on.

There are many different ways to generate metrics for your organization. You must ensure they are meaningful to you and your organization. Develop a plan for the types of metrics you want to gather. Also, ensure that the metrics you decide to gather are ones that the executive team and the BoD will want to know about. Vulnerability metrics are, therefore, important as they paint a picture of how you are mitigating risks in the organization.

Vulnerability metrics

The **National Vulnerability Database** (**NVD**) has created the **Common Vulnerability Scoring System** (**CVSS**). CVSS scores are then associated with the **Common Vulnerabilities and Exposures** (**CVE**) database to help the public understand the severity of a given vulnerability. Some vendors have also developed their own vulnerability scoring and reporting tools. So, how can you tell when a risk is high or low?

At the time of writing, the CVSS is on version 3.1 and takes in a number of metrics to determine the severity. Scores based upon the CVSS version 3.1 go from none to critical with their range of scores as follows:

Severity	Base Score Range
None	0.0
Low	0.1 – 3.9
Medium	4.0 – 6.9
High	7.0 – 8.9
Critical	9.0 – 10.0

Table 12.1 – CVSS scores

To score a vulnerability, the NVD uses a number of metrics to build its scoring model. For instance, it will score higher if the vulnerability requires user-level or administrative-level privileges to exploit. Does the vulnerability require any user interaction for it to be exploited? How complex is the attack to pull off? Needless to say, the easier it is to pull off an attack with little to no user interaction will result in a higher score.

You will need to build metrics for how you remediate these types of vulnerabilities. For example, many external regulators expect critical and high vulnerabilities to be remediated within 48 hours of being detected. Are you able to meet those expectations in a timely manner? How can you prove that you have met your or their requirements? You will need to track which vulnerabilities affected an IT resource and how quickly you and your team resolved that particular issue.

As an example, the Log4j vulnerability received a critical base score of 10 for CVE-2021-44228. This meant a perfect storm was created as there were a number of organizations that were exposed to this vulnerability. It also meant that it was easily exploitable. Metrics should be developed to show how you and your team quickly took action to eradicate that vulnerability from your environment. Those metrics can then be reviewed at a given interval to show how your **Security Incident Response Team** (**SIRT**) team responded to a given threat in the environment.

I, however, have a love-hate relationship with CVSS in general. While scoring a vulnerability may be the same across all CVEs that have been released, not all CVEs have the same risk exposure. For example, say you had the Heartbleed vulnerability present in your environment. As you may recall, Heartbleed is a vulnerability that affects legacy versions of SSL and TLS version 1.1 and lower. The CVSS score for the vulnerability (CVE-2014-0160) is 7.5, or a high ranking. The ability for someone to take control of a session externally is high. How risky is it, though, if it is only accessible internally and is behind three different firewalls?

Three different firewalls could be an exaggeration; however, it proves a point. An internal web application, vulnerable to Heartbleed, is less susceptible to risk than the same web application being exposed externally to the internet. Again, you could have roughly 4+ billion users browsing your website at any given time. That is a much larger risk than, say, 1,000 internal corporate users who may have access to the corporate website.

When reporting metrics on vulnerabilities, ensure that you are reporting them properly. You could split up your reporting of vulnerability risks based on the location of the IT resource, such as external or internal resources. That way, you can provide better metrics when reporting which vulnerabilities were mitigated and why. You would have a better chance of explaining a story of why part of your environment has all the critical and high levels remediated in the DMZ, while you may have a couple of highs in your internal network.

Poor **mean time to identify** (**MTTI**) and **mean time to contain** (**MTTC**) could result in poor cyber hygiene. You will need a vulnerability scanner in your environment to detect vulnerabilities in your IT resources. Your executive team and possibly the BoD will want to know how quickly a vulnerability was detected in the environment and how you responded and mitigated that vulnerability. You will also want to perform and report on assessments regarding your scanning footprint. Do you have agents installed on all your IT resources? Which ones are not? Are they all being scanned according to corporate standards?

The organization should also have documented **mean time to remediate** (**MTTR**). Remediation of vulnerabilities should fall under your standards documentation. This standard should lay out any **service level agreements** (**SLAs**) for when your stakeholders and customers expect them to be resolved. Remember, it is quite possible that you are not only holding on to your sensitive information, but you are also holding onto the sensitive data of others too. You must make sure that the systems are secured according to established policy documents and that they are remediated based on SLAs.

Another indicator to report on is whether the vulnerabilities were actually closed after remediation. Your security team should track which vulnerabilities were discovered and which ones have work orders for remediation and re-testing to validate that the remediation worked. These can be tracked in a number of different ways, either in the vulnerability management tool itself or your **information security management system** (**ISMS**). You should build metrics around how many work orders or tasks have been created versus how many have been remediated.

While vulnerability metrics can show levels of risk within the environment, incident reporting can help determine how your team reacts when an incident does arise. These can come in the form of red and blue team exercises. Team exercises are a great way to not only build muscle memory for how your team responds but also allow you to build metrics around how quickly the team responded to and mitigated a threat in the environment.

Incident reporting – red team versus blue team

Incident reporting can take on a number of different topics: those incidents that happened to you and your organization and those that were response exercises. Both topics should be reported to the executive staff and the BoD. You should disclose incidents when they occur using your escalation parameters based on the **Traffic Light Protocol** (**TLP**). Reporting incidents is also becoming a requirement if you are a publicly traded company in the U.S. This is due to the Securities and Exchange Commission making publicly traded companies become more transparent when it comes to security incidents and breaches.

We touched on the topic of **incident response** (**IR**) in *Chapter 7*; however, we only scratched the surface of how to test your IR plan. We briefly went over tabletop and live-action exercises, but how would you go about architecting one, and how would you build metrics from it? Next, we will cover red and blue teaming exercises and how metrics are built around them.

I am sure we are all familiar with what a red team is. If you have ever performed an ethical hack against your infrastructure or hired a penetration testing company to perform one for you, this is an example of red teaming. Most think of red teaming exercises as someone using a penetration testing toolkit such as Kali Linux or Metasploit to see what they can break into. Penetration testing is used as a final exam where you see how your hard work has paid off. However, when performing a penetration test, you should have a follow-up test scheduled 3-6 months after the initial test. This allows you to check whether your remediation steps have successfully closed any of the gaps or vulnerabilities identified. This is also the time when you should have your final report in hand, along with your remediation report to better understand your remediation efforts.

For instance, you received a report that says you have 2 criticals, 4 highs, 10 mediums, and a bunch of low or informational vulnerabilities. Once you have the final report in hand, you begin working on remediation steps. A period of 6 months goes by, and you have remediated most, if not all, of the identified findings in the final report. Now it is time to take the final report, along with any findings from the remediation report, and determine how well you did. This will also highlight any deficiencies you may have encountered during the remediation; for example, maybe there were not enough people or capital to fully remediate all the identified findings.

Blue teams refer to a group of individuals who are responsible for defending an environment from the red teams. They use the tools at their disposal to identify indicators of compromise across the network. They can use tools such as network sniffers, review security logs, vulnerability scanners, and firewalls to detect an imminent threat. Blue teams also defend their systems against threats and vulnerabilities that are thrown at them by the red team.

When reporting on blue team scenarios, you should build metrics based on how many detections were picked up, the number of attacks that were stopped, and how the blue team was able to block incoming attacks from the red team. The red team should build metrics based upon the opposite; how many attacks were successful and what types of vulnerabilities were discovered.

As the head of security, it is also your responsibility to track actual threats as well. Exercises are great to report on as they help tell the story that the SIRT team is being prepared for when something does happen. When it does, you and your team shall be prepared for incidents as they arise. You should also maintain historical reports based on actual events. These could be lost or stolen laptops or desktops, account takeovers, **Distributed Denial of Service (DDoS)** attacks, or any other type of incident you may encounter. As you build out these metrics, ensure that they can be stored somewhere for historical purposes so you can see how you and the organization have progressed over time.

Reporting to the executive team and BoD

Once you have collected all the metrics necessary to show the progress of your security program, it is time to report it. Again, you must know your audience and correlate metrics data to a format that the executive team will understand. The next three sections will provide insight into how to do that.

System security plans and the risk register

System security plans (SSPs) can be used for more than just recording security configurations of an IT resource; they can also be used as part of your metrics reporting. As part of the SSP, a risk register should also be created to record vulnerabilities and lacking security configurations. The risk register, much like the SSP, is a living document for the IT resource and should be kept up to date when new risks arise.

When reporting metrics from an SSP and its associated risk register, we must first identify what those risks are. Risk registers can come in many different forms and capture plenty of different metrics; however, to make it meaningful, you should at least capture what resources will be needed. Resources, such as capital and operational expenditures, which include time and materials, must be captured. When reporting on the risk registers, we can use a balanced scorecard to display progress and expenditures. An example of a risk register is shown in the following table:

ID	Risk	Risk Score	Response	Cost	Time	Owner	Status
1	Log4j	25	Mitigate	$5,000	4 wks	Jon Snow	Open
2	Old Tomcat version	25	Mitigate	$0	2 wks	Jon Snow	Open
3	Old Java version	20	Mitigate	$0	2 wks	Jon Snow	Open
4	End of life operating system	9	Mitigate	$1,500	2 days	Robb Stark	Closed
5	Hosting	6	Transfer → AWS	$7,000	2 mnths	Sansa Stark	Closed

Table 12.2 – Risk register

As you continue to identify risks in your IT resources, you continue to add to the register. The closed items in the register never get deleted; they stay in the *Closed* state. Reporting on these projects provides valuable metrics for your executive staff and the BoD. There are a number of different ways that you can report on them as well. For instance, risk heat maps and balanced scorecards are additional ways of presenting this data.

A risk heat map

One way to display the risks to your IT resources is through a **risk heat map**. By taking the risk score shown in the following table, we align the score to the **Impact** and **Likelihood** of the risk map. The risk score can be a combination of outside risk factors, such as CVSS, placement of the IT resource, mitigating controls, and the type of data processed, stored, or transferred on the resource.

Impact	Critical	5	10	15	20	25
	High	4	8	12	16	20
	Moderate	3	6	9	12	15
	Low	2	4	6	8	10
	Rare	1	2	3	4	5
		Rare	Low	Moderate	High	Critical
	Likelihood					

Table 12.3 – A risk heat map

In this example, we have an Apache Tomcat server running an older version of the Red Hat Enterprise Linux operating system and it is susceptible to Log4j. The server is externally facing, meaning it is customer facing, and stores PII data. Due to the CVSS score of the vulnerability, it is rated as a high-risk resource that should be fixed as soon as possible. By aligning the risk scores to the heat map, you can present the risks involved to executive staff and the BoD.

To calculate the overall risk score, we use the following formula:

risk = likelihood x impact

In *Table 12.2*, we have a risk score of 25 for Log4j. To calculate this, we first take the impact score ranging from 1-5 and the likelihood score also ranging from 1-5. We multiply the two scores together to get the overall score of the identified risk.

Balanced scorecard

When presenting the information to executives and the BoD, you must remember that IT and security may not be their areas of expertise. This can hinder your ability to communicate certain aspects of your security program. In other words, you must get to know your audience and how to speak to them effectively without getting into the weeds of a piece of technology. This is where your stories are used to define the metrics you have built.

Depicting this information can also be challenging when you do not know your audience. Overgeneralizing or simplifying your message will be an important key to delivering your metrics. Business leaders want to see these metrics in charts or graphs, visualizations that are easy to read and understand. A **balanced scorecard** (**BSC**) is one way of displaying all your hard work in a single graph. While a BSC was originally meant for for-profit organizations to display outcomes of business objectives, it can also be used to display security initiatives.

When using a BSC to report on your security program, you can use a number of different topics. However, when simplifying the approach of reporting to the executive team, you should report on the following key pieces of information:

- Project objectives and timelines
- Project expenditures
- Resources, both CapEx and OpEx
- Incidents and SLAs

These are not the only areas of focus for your BSC; however, keep in mind that you are now presenting business-related topics to those outside of IT. They will not care how many vulnerabilities you have remediated, nor will they know what a **web application firewall** (**WAF**) does. The question you are trying to answer is whether the organization is secure or not. While this question may seem open ended, this is what they want to know. You must present where the security program is at with the tools that you were given.

Summary

You are responsible for the security program for your organization. It is up to you to highlight all the hard work that you and your team have put in. A large part of telling this story is based on the metrics you have collected over the months since your last report, though that is only part of the equation. You must have a story lined up to help sell all the hard work. Without a story, that is all a report is: just numbers.

Metrics should be based on mitigating risks in the organization. These risks include what is internal to your company and external for both third and fourth parties. When reporting on internal metrics, pay close attention to business goals and objectives along with external regulatory requirements. Build metrics around project outcomes and leftover tasks which still need completion. Obtaining PCI or ISO certifications can also build a story around your security program and how you align against your competition.

Lastly, represent your findings in simple ways that those outside of IT and security can understand. Not everyone you present to has an IT or security background. Get to know your audience before presenting to understand better who you will be speaking with. You will want to present business-type goals to your executive team, including finances and project outcomes.

I would like to leave you with a few parting thoughts as we draw the book to a close.

Building relationships and developing *soft* skills will be important as you move from an engineering/architectural role to the head of security. You will need to understand the business objectives and align those to the security program you are developing. Taking these initial steps will only make you more proficient in your position. As you build these relationships, remember that you will also need to sell your ideas to the executive management and the BoD. You will need confidence when you sell your ideas for how you believe what you and your team are doing will put the organization in a better position.

Soft skills also require that you have a sense of empathy when it comes to dealing with coworkers. It is inevitable that you will experience a cyber incident, and it will affect those around you. While your day may have been impacted by the incident, nothing compares to the impact it had on a coworker. Again, whether it's one person or an entire department, know your audience and reassure them that it will be taken care of.

Thank you for taking the time to read this book. I really appreciate it. If you have questions, feel free to reach out.

Email: jason@jasonbrown.us

Website: https://jasonbrown.us

Twitter: @jasonbrown17

References

1.   NIST CSF scoring template: https://github.com/PacktPublishing/Executive-s-Cybersecurity-Program-Handbook

2.   BitSight risk analytics: https://bitsight.com

3.   Security scorecard, security ratings, and risk analytics: https://securityscorecard.com

Index

U

V

W

Z

Packtpub.com

Subscribe to our online digital library for full access to over 7,000 books and videos, as well as industry leading tools to help you plan your personal development and advance your career. For more information, please visit our website.

Why subscribe?

- Spend less time learning and more time coding with practical eBooks and Videos from over 4,000 industry professionals

- Improve your learning with Skill Plans built especially for you

- Get a free eBook or video every month

- Fully searchable for easy access to vital information

- Copy and paste, print, and bookmark content

Did you know that Packt offers eBook versions of every book published, with PDF and ePub files available? You can upgrade to the eBook version at packtpub.com and as a print book customer, you are entitled to a discount on the eBook copy. Get in touch with us at customercare@packtpub.com for more details.

At www.packtpub.com, you can also read a collection of free technical articles, sign up for a range of free newsletters, and receive exclusive discounts and offers on Packt books and eBooks.

Other Books You May Enjoy

If you enjoyed this book, you may be interested in these other books by Packt:

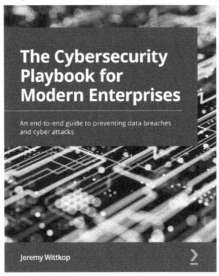

The Cybersecurity Playbook for Modern Enterprises

Jeremy Wittkop

ISBN: 9781803248639

- Understand the macro-implications of cyber attacks
- Identify malicious users and prevent harm to your organization
- Find out how ransomware attacks take place
- Work with emerging techniques for improving security profiles
- Explore identity and access management and endpoint security
- Get to grips with building advanced automation models
- Build effective training programs to protect against hacking techniques
- Discover best practices to help you and your family stay safe online

Hack the Cybersecurity Interview

Ken Underhill, Christophe Foulon, Tia Hopkins

ISBN: 9781801816632

- Understand the most common and important cybersecurity roles
- Focus on interview preparation for key cybersecurity areas
- Identify how to answer important behavioral questions
- Become well versed in the technical side of the interview
- Grasp key cybersecurity role-based questions and their answers
- Develop confidence and handle stress like a pro

Packt is searching for authors like you

If you're interested in becoming an author for Packt, please visit authors.packtpub.com and apply today. We have worked with thousands of developers and tech professionals, just like you, to help them share their insight with the global tech community. You can make a general application, apply for a specific hot topic that we are recruiting an author for, or submit your own idea.

Share Your Thoughts

Now you've finished *Executive's Cybersecurity Program Handbook*, we'd love to hear your thoughts! Scan the QR code below to go straight to the Amazon review page for this book and share your feedback or leave a review on the site that you purchased it from.

https://packt.link/r/180461923X

Your review is important to us and the tech community and will help us make sure we're delivering excellent quality content.

Download a free PDF copy of this book

Thanks for purchasing this book!

Do you like to read on the go but are unable to carry your print books everywhere?

Is your eBook purchase not compatible with the device of your choice?

Don't worry, now with every Packt book you get a DRM-free PDF version of that book at no cost.

Read anywhere, any place, on any device. Search, copy, and paste code from your favorite technical books directly into your application.

The perks don't stop there, you can get exclusive access to discounts, newsletters, and great free content in your inbox daily

Follow these simple steps to get the benefits:

1. Scan the QR code or visit the link below

https://packt.link/free-ebook/9781804619230

2. Submit your proof of purchase
3. That's it! We'll send your free PDF and other benefits to your email directly

www.ingramcontent.com/pod-product-compliance
Lightning Source LLC
Chambersburg PA
CBHW060549060326
40690CB00017B/3658

* 9 7 8 1 8 0 4 6 1 9 2 3 0 *